Life After
Johnnie Cochran

life after
Johnnie

Cochran

Why I Left the Sweetest-Talking, Most Successful Black Lawyer in L.A.

BARBARA COCHRAN BERRY

with JOANNE PARRENT

BasicBooks
A Division of HarperCollins*Publishers*

*This book is lovingly dedicated to both of my beautiful daughters,
Melodie and Tiffany, from whom I kept all of this buried
deep inside of me for many years.*

*All of us loved, inspired, and nurtured each other during our
growing years. May God always bless and keep them.*

With much love, respect, and admiration,

*Your mother,
Barbara*

Acknowledgments

We would first like to thank each other. In working on this book we've had the truly fortunate experience of sharing much laughter, many tears, and a few headaches under the pressure of it all, and ending up after these intense months each very fond of the other and happy to have found a new friend.

Next, we'd like to thank Brenda Feigen, our outstanding agent. She saw the potential in this book, first introduced us, and placed this book with a house and an editor who understood the message for women we wanted the book to convey. Her feedback on the content and structure of the book was also enormously helpful.

That brings us to our hard-working, enthusiastic, and delightful editor, Susan Rabiner. She believed in us and the book from the beginning. She added a special dimension to the project with her intelligent editing and her always gentle prodding. We knew we were in good hands with Susan.

We would also like to thank Gloria Allred, who first referred Barbara to attorney Michael Donaldson. It was Michael who brought in his office partner, Brenda Feigen. We are grateful to Gloria Steinem for giving us encouragement and reading some early chapters of the book. Barbara would also like to thank Juanita Walker Deckard, who helped jog her memory, and Monica Gunning, who graciously read chapters of the book. Finally, Mary Low from "Nine to Five" did a great job transcribing our many hours of interviews.

Barbara Cochran Berry
Joanne Parrent

Introduction

It was one of those beautiful southern California days—bright, sunny, glorious—the kind that leaves you glad to be alive. I had such a reputation as a Pollyanna while growing up that my friends used to refer to a day like this as a Barbara Berry day, because it filled everyone with hope and optimism. Perfect for thinking favorite thoughts—about my daughters, my job, my life—nothing serious or even remotely sad. Little did I know that before the sun went down, I would be forced to relive hurtful old wounds from almost twenty years before, wounds I had long ago consigned to the darkest corners of my memory. That day was also the beginning of a series of events that led to this book.

It started with a phone call from Johnnie Cochran, lead counsel of O. J. Simpson's "Dream Team" of defense lawyers. Upon hearing his voice my first thought was, "Well, well, well, the ex-husband from hell." On this particular day, his affected smoothness was in Hall-of-Fame form. He oozed so much cordiality that I could barely stand to keep the phone receiver against my ear. "Hello, darling. How are you today, dear? This trial is really keeping me so busy, sweetheart." He did several minutes of his best charm, the kind I remember his turning on whenever he wanted a special favor. Sad to say, human nature being what it is, it usually worked for him. You have to have been married to him and have seen him charm so many others so successfully. You have to have heard him tell stories of having charmed others to the point that their heads bobbed in inane nods, muttering "yes, yes," to whatever he said, to recognize the skilled routine for what it is.

Finally he came to the point: "Barbara, a reporter has opened up our divorce records. There was a sheet that had some bad things on it about me." I listened, wondering where he was leading. He continued, shifting smoothly from the smoke-blowing mode to the husband-knows-best mode. Everything would be fine, he instructed, if I'd "just deny the allegations." I frowned to myself, but did not challenge him directly. It was too nice a day to get into a test of wills with this iron-willed man. I just said, "We'll see, John." Confident he had adequately briefed me, John said he would give the reporter my phone number and then hung up.

A couple of days later Michael Goodman called. A reporter working on a cover story about John for the *Los Angeles Times* Magazine, he had gotten a copy of our divorce records and was calling because John had told him that I would deny the allegations of physical abuse described in the court papers. It had been a long time. I couldn't remember what those papers said. I asked him, "Just what am I supposed to deny?" He read from a declaration I had filed in May 1967, to obtain a restraining order against John. It said:

> On April 29, 1967, my husband violently pushed me against the wall, held me there and grabbed my chin. He has slapped me in the past, torn a dress off me [and] threatened on numerous occasions to beat me up.

From a second document, filed ten years later, he read:

> At the family residence on Sutro Avenue, [Cochran], without any reasonable cause, provocation or justification, physically struck, beat and inflicted severe injury upon the person of the Petitioner. Throughout the period of this marriage, he has continued to verbally harass [me] ... mindful of [Cochran's] tendency toward violence I fear that unless enjoined and restrained by appropriate Court Order, he will inflict injury on [me] and will continue to harass and annoy [me].

As Goodman read to me, vivid memories of the cruelty, abuse, and violence I had suffered during those years filled my mind. My strong reaction surprised me. After all, it had been eighteen years since my divorce from Johnnie Cochran. A long moment passed before I responded. "I am not denying anything," I said to the waiting reporter. He was shocked. John had assured him that I would take back everything.

Once the reporter hung up, I sat there and fumed. John had simply assumed that if he told me to lie for him, I would nod and do it. But I was out of the habit of wifely compliance and had been for years. When the reporter confronted John with the allegations, why didn't John simply admit that they were true? He could have said that he regretted what had happened, that it was a long time ago, and that he had grown as a person since that time. In the daily blitz of newsbreaks coming out of the O. J. business the whole incident might have blown over in a few days as reporters chased new revelations about the case or other participants. But taking responsibility for his actions has never been John's style—indeed, knowing how to avoid doing so may be one reason why he's so skillful at helping his often guilty clients escape responsibility for their acts.

For my own part, I was angry at his arrogance in believing that so many years after our divorce I would have grown in independence so little that I would mindlessly agree to play whatever role he chose to assign me. In that moment, I vowed to myself never again to be an accomplice in his well-orchestrated self-promotion schemes.

John called again a day or so later. Giving me my orders hadn't worked, so this time he was bargaining. "Barbara, you will want for nothing, ever, if you'll just deny the allegations. Tell the reporter I was a wonderful guy and that will be the end of the whole thing." Was this a bribe I was being offered? What happened to the search for the truth, what John claims he's so concerned about in the courtroom? Would he stop at nothing to create the picture of reality he wanted the world to see? Thinking back on it, I'm sure he so much wanted to come out of the Simpson trial and the rash of publicity surrounding it like a squeaky-clean defender of truth, justice, and an American hero wrongly accused that if I had been prepared to deny those allegations I could have negotiated for much more than I received for this book. But he was talking to the wrong wife. If he was going to repackage the reality of his own life, he was going to have to do it without the connivance of one of his victims.

As he waited for my answer, pressuring me with his silence, I gave him the news he didn't want to hear: "I am not going to deny anything." There was a dead silence. Anyone who knows Johnnie knows it takes a really big shock to leave him without his next line. After a long silence, he retorted, "I will never ask you to do anything for me again." His voice was full of anger, but I was long past the time when I automatically felt frightened whenever he was angry. I said that was perfectly fine with me, and we both hung up.

John is not the type to give up easily, and I waited for his next move. When it came it involved one of his big guns—his father, Johnnie Sr. John's father and I have always had a beautiful friendship, one that has lasted over three decades. The "doctor," as I call him, had always been much more than a father-in-law to me.

He was a kind friend. On the Sunday morning that the article about Johnnie appeared in the *Los Angeles Times*—complete with my refusal to deny the allegations of abuse—the doctor called. I don't know why I was surprised when I heard his voice. He greeted me cheerfully, "Good morning, Bobbie"—he always called me Bobbie. I responded enthusiastically, "Good morning, doctor." Soon his tone changed, "Did you see the article in the paper this morning?" I told him I had seen it. There was an awkward silence. He remarked that the reporter had really tried to write something that would sell. I responded that, yes, he certainly had a winner. Then, more firmly, more fatherly, he got to the point of the call. "Other reporters are going to be coming around now, so just deny everything. Tell them that everything was blown out of proportion." I mumbled something vague. We wished each other a great day and hung up.

As I put down the phone, a wave of sadness came over me—I felt the pain of losing a friend. I knew that I was going to have to disappoint that strong, staunch elder that I so respected. The days were gone forever when I would lie for his son, but it was not because I wanted to hurt the man I had once been married to. It was just that I believe strongly that in order for a person who has ever been abused to stop seeing herself as a victim, she must stop denying that the abuse occurred. Once you say it happened, and know you have taken steps to end it, it becomes the problem of the abuser. As long as you say—to yourself or to others—it did not happen, it remains your problem.

The doctor was right. It wasn't long before the press was after me, in the usual feeding frenzy over anything related to the O. J. Simpson trial. I came home from my teaching job the next afternoon, planning to change into exercise clothes and go off to a step class at my gym. But before I could change I happened to glance out the window and notice a large TV van pull up in front of my house. I decided not to answer the door—the blinds were all closed—and stay in for the evening. It was after midnight before the television crew finally left. But I knew that although I had avoided them once, they weren't going to give up.

Sure enough, in the days that followed, reporters chased me around, hounded me, and camped in front of my home. John, meanwhile, began to prance around frantically, making the rounds with his current wife, Dale, from one TV station to another. I thought he must have hired a helicopter to get him from one studio interview to the next. Within one week, he was on *Current Affair*, *Day One*, *Prime Time Live*, and the local news, still vehemently denying that any abuse had taken place. To someone who knew the man, the entire spectacle was part of a skilled plan for spin control. But as I watched him work the public, I had a new feeling as well. I had faced the truth about our relationship and now it might do him some good to do the same. Over the nearly two decades we were apart I had never come after him demanding that he admit what he had done to me and apologize. Rather, it was he who had come to me asking that I deny the truth about it. When that failed, he was now trying to convince the world that I had not spoken truthfully. It was time for me to tell the truth and let the public hear my side. But to do an honest job of it, I had to tell the story in its full context, and in telling this story, I would have to paint a full portrait of John as he was during our years together—a man of much charm and many talents, but as well a man of serious flaws.

I made a simple, short statement for the reporters. I said that I was happy for John's phenomenal success and that I had no comment at this time. I intended to write my story in a book.

When John heard that I planned to write a book, he was beside himself. But, remember, he had said he would never ask me for anything again. Shackled by his own ultimatum, he now couldn't call me directly and try to change my mind with new threats or bribe offers. So he called one of our daughters and tried to get her to intervene, telling her, "You know, I will sue your mom if she writes a book." She was smart enough to inform him that she would not get involved in the situation.

I have recently thought deeply about why I had kept quiet about my life with John for so many years. The reasons were the same as those many women have for not talking about the abuse in their marriages. I didn't want our children hurt, humiliated, and embarrassed while they were growing up. I wanted our two daughters to love and respect their father, despite how he treated me. But now they are both adults and I no longer need to protect them. They have built their own relationships with him and can separate how he is as a father from how he was to me as a husband.

I had also kept things hidden because I felt that I didn't want to share my innermost thoughts with the world. Unlike John, I'm a private person. But as I thought about that reason for keeping secrets, I'm now sure that shame played a role. I didn't want to admit to myself that I had been treated so badly by a loved one— by someone who claimed to love me—for what might that suggest about the kind of person who would take abuse from a lover? Ironically, it may have been John's call to me that forced me to face the truth. Had he said nothing to me, and had the reporter caught me off guard, I might have given out a story John would have been happier with—maybe that there are problems in all marriages, but that it was now all ancient history and not worth talking about. But John's attempts to involve me in a cover-up provoked a feeling that I would not be serving my own needs, or those of other abused spouses, by helping an abuser make light of a history of abuse.

In fact, my decision to be openly honest about my own abusive marriage led to the most compelling reason to write this book— to send an encouraging message to other women who are still involved in abusive, humiliating, or intolerable situations. After all, I did manage to break free. While it's necessary for abused spouses to know that they are not alone, that others are experiencing precisely what they are experiencing, it is also important that they know that there is a way out, and that others have found the strength to do whatever was necessary to get themselves through that door. If only one woman finds in my story the strength to extricate herself from a dangerous or degrading situation, then it will have been more than worth the pain of opening old wounds.

In the days after the uproar over the *Los Angeles Times* article, while weighing the pros and cons of writing this book, I spoke to my minister. He told me, "We become fulfilled by emptying ourselves out." I decided that now was the time to tell the truth—to empty myself out.

What follows is my story—the true story of my life with a famous attorney, a man who in his public life is seen as successful, accomplished, bright, and charismatic, but who revealed in our life together a private side that was often deceitful, manipulative, controlling, and abusive. To the extent that these traits go hand in hand with being a successful criminal defense lawyer, that is an element in this story. And, as the Simpson case has so tragically illuminated, to the extent that people may be more willing to condone these abhorrent characteristics in wealthy or prominent men, often encouraging their wives to put up with humiliation and abuse far longer than if they were married to less powerful men, that too is a very real part of this story.

PART I

Barbara Berry

Chapter 1

In the early 1930s, the country was caught in the devastating clutches of the Great Depression. In the South, where I was born—in Shreveport, Louisiana—things were especially bleak. And they were worse than bleak for black people. Eartie Norris, my mother, was the second youngest of a family of ten children from Grand Cane, a rural town just outside Shreveport. Eartie was a bright girl, but Grand Cane didn't have a high school. To finish her schooling, Eartie had to make the decision to move to Shreveport and live with relatives. She was the only one in her family to do so. Eartie loved school, and secretly hoped for a miracle—one that would bring her the money to go to college so that she could become a teacher. But no miracles came and no money. Eartie's parents and her older brothers and sisters were barely getting by; the only source of work—work that paid next to nothing—was in the cotton and sugar cane fields that surrounded the area. Supporting an able-bodied young adult for four more years would have been a real burden. College tuition was out of the question.

What did come along for Eartie, however, was the sweetest, best-looking boy she had ever seen. Youree Berry, from another small town outside of Shreveport—Stonewall, Louisiana—had also come to the city to go to high school. If opposites attract, this was a classic example. While Eartie was reserved and shy, like an unopened flower, Youree had an exuberant, outgoing personality. She loved his storytelling and his lively sense of humor. He loved her quiet intelligence and her compassion. They married right after they graduated from high school and took up the task of trying to support themselves as best they could in this southern city that prided itself on being the Confederate capital of Louisiana. All Youree could manage to find was a job as a bicycle-riding delivery boy for a drug store. As he wheeled around the streets of the city delivering prescriptions, he knew he wanted more for himself and his young wife, but wasn't sure how or where to find it. Eartie, meanwhile, contributed to the young couple's finances by taking in the ironing of well-to-do white people.

Before long, Eartie was pregnant. She gave birth to a baby girl. One evening, shortly after she came home from the hospital, the proud parents put their baby daughter to bed in her crib. When they next went in to check on her, she was dead.

When Eartie became pregnant again, she and Youree were afraid to invest too much hope, or make too grand plans for this next baby. They still hadn't been able to sit down and pick out a name when Eartie went into labor and entered Charity Hospital. After a baby girl was born, the proud father brought Eartie a movie magazine to read while she was recovering in the hospital. On the cover was Eartie's favorite actress, Barbara Stanwyck, and that was it. Eartie said to Youree, "Let's name the baby Barbara Jean."

I doubt that I was as outspoken as a child as I am now or I might have announced: "How dare you name me Barbara Jean." Every third black girl my age seemed to be named Barbara Jean. It wasn't much consolation that every little white girl was named Barbara Ann. Could Barbara Stanwyck have been that popular? But, like it or not, there I was, unleashed on the world, little Barbara Jean Berry.

Eartie and Youree showered me with love and affection. I was their precious child, the one who made it past those first few weeks, and they didn't want anything bad ever to happen to me. Little did they know that something did happen right there in

4

Shreveport about a year and a half after I was born, something they wouldn't be able to protect me from. Unbeknownst to me, as I was toddling around my parent's home, another black child was born in Shreveport. This child would grow up to be Johnnie L. Cochran, Jr. But that's for later. Now, I need to tell you how I became the young woman who met him two thousand miles from the place of our birth and later married him.

At that time, Youree was still pondering the problem of how not to be a delivery boy all his life, and Eartie, now with me underfoot, hadn't grown any fonder of ironing other people's wash. One day, when I was three, Eartie's Uncle Riley came to visit us. Uncle Riley had grown up in Grand Cane, but now lived in San Francisco. He had a good job as a Pullman car porter for the Southern Pacific Railroad. On that eventful visit, he regaled my father with stories of life in the West. Youree listened, enthralled and excited, as Uncle Riley advised him, "There are all kinds of jobs, Youree. Come out to California." Of course, Youree likely knew that it was an exaggeration to say at the time that there were *all* kinds of jobs available to a black man, but he would have been happy if there were at least some jobs a grown man could do with dignity.

Youree and Eartie made a plan, a way to escape. He would go to northern California and find a job and a place to live and then send for Eartie and me. It wasn't long before we were on our way to Oakland, California. My parents were thrilled to be there, but my first impression of Oakland was of its climate—very different from the sunny South. I watched—amazed at first, but then just dismayed—as day after day the fog rolled in through the early evening, stayed all night and the next morning, then lifted for a few brief hours in the afternoon. But the family hope and excitement in our new life quickly overshadowed my consternation with the dreary weather.

We moved to a block of look-alike houses, old two-story bungalows that were built so close together that you could put your head out your window and talk to your neighbor as comfortably as if you were inside the same house. Most of the houses on the block were painted an ugly dark color—to match the weather, I presumed. A mom-and-pop store was on almost every corner. There were no security bars on the windows, very few fences, and people usually left their front doors wide open. It was a safe and comfortable place to grow up.

Shortly after we moved, Youree landed a very good job. His Uncle Clyde, who had moved to Oakland sometime before us, had started a dry-cleaning business. Clyde told Youree he needed help and Youree soon went to work at the dry cleaners. He worked long hours but was paid well—we even got free dry cleaning.

As I was growing up, Eartie and I were constant companions. We went everywhere together—the park, the library, the movies, church, the market. We were like sisters, or best friends. I appreciated that she never talked to me as many adults talk to their children. Nor did she or my father ever spank me. Later in life when my husband hit me it was almost unreal, like something from a bad soap opera. There had been nothing in my childhood to prepare me for it.

Every day, my father drove me to my elementary school, which was near my uncle's dry cleaners. It was an integrated school, mostly white with a few black children—something that could not have existed in Shreveport at that time. When I first started school, my teacher and the principal called my mother in for a conference. I was worried sick that I had done something wrong. It turned out that they thought I was an excellent reader and suggested that I skip the first grade. So I did. I was still the best reader in second grade, but I had a lot of trouble mastering the way they wanted us to write. I couldn't do it. My mother took on the task of teaching me handwriting. Doing so confirmed that she would have been an excellent teacher—it wasn't long before I was getting penmanship certificates.

Eartie often talked about how much she wanted me to be able to go to college one day. She told me how she hadn't been able to become a teacher because her family hadn't had enough money for college. But she clearly hadn't lost her desire to learn. When I came home from school each day, she asked me about everything I was studying—she was fascinated—it was so much more than she had been exposed to in her own schooling in the South. We would sit in our high-ceilinged living room with its dark burgundy velvet couch and matching chairs and talk for hours. I'd put my books down on the nondescript coffee table with family pictures under the glass top or on the end tables with stiff doilies, frilly lamps, and more family pictures. Across the room was music-loving Youree's big, old-fashioned brown radio and a record player, with stacks of records on the floor under it. I loved that dreary-looking room not because of the decor but because of the good times I had there, the talks, the memories I still have of it.

My teachers all through school were white, female, and usually old and frumpy. It seemed like they all wore the same dress every day and had been there since the beginning of time. There was one exception. In the fourth grade we had Mrs. Dyer. When we saw her, we went crazy. She had blue eyes, blonde hair, and was young and hip. I guess if I ever had a teacher who was a role model for me, it was Mrs. Dyer.

My father was making a good living at the cleaners now. We were doing fine. But my mother didn't forget about her relatives back in Louisiana. She sent them my hand-me-down clothes and several newly invented gadgets. Thanks to Eartie and Youree, my Grand Cane relatives got their first electric iron and later the first electric refrigerator in that little town, complete with the money to install it.

Most summers, my mother and I went back to Louisiana to visit. I loved the pine trees and the green sugar cane fields, but noticed, even as a young child, that countless black folks, including many of my own relatives, still worked in back-breaking jobs to maintain themselves in spirit-breaking poverty.

Eartie was a God-fearing Southern Baptist—a very religious one. She was a Sunday school teacher at our church, fulfilling her old desire to teach. Youree had been brought up as a Methodist, but came along to the Baptist Church with us so we could all be together. Despite listening to Baptist preachers Sunday after Sunday, Youree never took on Eartie's strict Baptist ways. He did, however, soon perfect an Oscar-caliber imitation of a Baptist preacher. I would fall down laughing when he started in with "Brethren, let us heal ..." He'd do all the "Yeas" and "Amens" from the congregation too. He'd carry on until Eartie would tell him he ought to be ashamed of himself. Then he'd stop. He loved to tease her, but he also respected her beliefs, and each appreciated the limits the other wanted respected.

Eartie didn't allow me to wear makeup (I was too young) or play cards (it wasn't ladylike). Of course, by the time I reached high school, everyone was wearing lipstick. I had peers to deal with and had to do something. I stealthily purchased something called Revlon Russian Sable. I put it on my lips the minute I got to the bus stop and rubbed it off on the way home. My lips must have been especially pink from all the rubbing, but Eartie apparently thought that was their natural color, because she never said a word.

When I went somewhere with Youree, I would put on my makeup. He didn't care. He was my buddy. We were always in cahoots, hiding things from Eartie that she didn't approve of. Like playing cards. Youree taught me bid whisk and bridge when Eartie wasn't home, gleefully liberating the hidden deck of cards.

My maternal grandmother, Patsy, in Grand Cane, was even stricter than my mother. She wouldn't let me wear my pedal pushers, the height of fashion at the time. She'd frown and say, "You don't wear pants here. Go in and put on a dress." During one visit, Youree and I hid under a tree behind the house to play some cards. My grandmother started out toward us, surprising us, and Youree, in a panic, threw the whole deck of cards up into the tree. The cards all sort of sat there, precariously balanced on the limbs and leaves of the tree, while we stood and smiled and said something dumb like, "Hi, Grandma Patsy. It sure is pretty out here today." By the time the cards started to fall to the ground, she—thankfully—had turned and gone back to the house. It was a close call. Youree spent the rest of the afternoon fishing those cards out of that tree.

Sometimes, on the way back from these visits to Louisiana, we'd stop in Los Angeles, where my father's mother, whom we called Mama Essie, and his younger sister, my Auntie Annette, had moved. They were the opposite of my mother's family, but they loved my mother and she loved them. Mama Essie was a young, rip-roaring type of grandmother. She liked to party and she loved to go to the horse races. While my mother wouldn't wear makeup or do much else to improve her natural good looks, Auntie Annette was not only attractive, she was stylish. A hairdresser, she knew how to put on makeup professionally. Every strand of her hair was always perfectly in place. And she dressed like a fashion model. Often my mother would let me stay on alone for a week or so at Auntie's house so I could play with my cousin, Doris, Auntie's only child, who was just a year older than I was. They were wonderful times with my fun-loving relatives, Auntie and Doris and Mama Essie, and I loved hot, sunny Los Angeles.

Then I'd return to dreary Oakland where every day was a bad hair day. When I started to straighten my hair, I'd work hard pressing my hair in the morning, but within minutes of stepping out into the murky Oakland fog, it was bigger and curlier than before I started.

These were the kinds of serious problems I had growing up. The weather you've heard about. There were picnics—I hated picnics. I didn't understand why we had to drive fifty miles to eat cold chicken in the sand, with ants crawling all over us. And then, there was the crisis over giving blood. The only time I remember crying during those years was one afternoon when they wouldn't let me give blood at school because I didn't weigh enough. I was devastated—everybody who gave blood got to go home early on Fridays.

When I got home that tragic day, however, I cheered up quickly. Friday was the night Youree and Eartie used to cook dinner together. It was fun watching them in the kitchen, laughing, talking, and working together, affectionate and playful. We'd all sit down to eat dinner together and talk about what happened that day. There was a great feeling of camaraderie. Even when Youree and Eartie disagreed about something, I watched them talk things out, each listening to the other's position. This was the picture of family life I took into my first marriage. But after a few years with John, I felt as though I had stepped into the wrong family photo.

Youree was grateful for his job at Uncle Clyde's dry cleaners, but he still had a nagging feeling of restlessness. He wanted to do more with his life. Every chance he got, he'd have our old brown radio playing, listening to the disc jockeys and the music. After church on Sundays, we'd often go to San Francisco to visit Uncle Riley and his family. One day Youree asked Uncle Riley for advice on how to go about trying to become a disc jockey. He was sure he would be a good one. Uncle Riley encouraged him to just go in and talk to the people at the radio station: "Tell them you can do it. You never know. Maybe they'll give you a chance."

I don't know how long Youree had been dreaming about being a disc jockey or how much planning and practicing he had been doing before this, but when I was eleven he came home for dinner one night and told us, "I want you two to listen to KSAN at 9 p.m. tonight." I asked him, "What's happening on KSAN at 9 p.m.?" He would say only that he wanted us to listen to something and see how we liked it. I told him he better find the right station for us because I never listened to KSAN. He went over to the ugly little brown radio that we had, set it to the station, and then left.

That night at 9 p.m., Eartie and I gathered around the radio. We couldn't believe what we heard. There was Youree's voice coming from the radio. He was doing a show, playing music, doing the ads and the news. And it was a good show. That night, Youree became the first black disc jockey in the San Francisco Bay area. He began doing the 9 p.m. show every night and soon the station gave him an afternoon show as well.

It wasn't long before Youree was a popular disc jockey in the area. He played mostly jazz—musicians like Duke Ellington, Les Brown, Count Basie—but sometimes he'd dedicate a rock-and-roll song to me just before he went off the air at night. He enchanted his listeners with stories about the great jazz musicians, stories he would weave throughout the show. His special talent, though, was his ability to create a mood that was perfect for whatever was happening in the world at the time. Like a therapist, he could connect with people and cheer them up or soothe them with the music he played.

Youree would often pick me up after school and I would go with him on his rounds visiting sponsors. Sometimes that would take us into a bar. We'd both have a Coke and sit around and talk with the sponsor. I reveled in the idea that I was probably the only twelve-year-old in the country who could hang out in bars. It would have driven my mother to distraction had she known. The sponsors gave Youree all kinds of things—records, record players, televisions. Sometimes I would stay with Youree at the station to watch him prepare for his show. I learned every detail about how to do a radio show and soon decided that I was going to be the first black female disc jockey. Youree promised to help me. I knew my mother wanted me to be a teacher. But I figured she'd have to settle for me going to college, while I also did my own radio program. I was a girl on top of the world then, thrilled about what life had in store for me.

When I was about fifteen, Mama Essie divorced her second husband—Youree's father had died when he was only a boy. After the divorce, she decided she wanted to get away for a while and moved from Los Angeles up to Oakland to live with us. I never understood why she would want to leave L.A. for this gloomy city, but it later turned out to be a godsend. At first, she stayed in my room, but things were going so well that we moved to a bigger house in East Oakland where we each had our own room again.

I got good grades in high school, and also managed to be part of the social activities and clubs. I dated a few boys—no one special. Eartie, of course, was very watchful of my social life. She didn't want anything to get in the way of my going to college— particularly marriage or children. This was an area where her Baptist strictness really came into play, putting a real drag on my dating and party life. Eartie insisted that I be home by 11 p.m., even though *everyone* else could stay out until midnight. When I did go to a party, I'd try to unobtrusively make my exit at 10:45 and slip into my father's waiting car, hoping the others wouldn't notice that I was leaving so early.

Also on the subject of dating, Eartie would always say to me, "What man will buy a pair of old shoes if he can get a new pair?" The implication here was that I would be like a pair of old shoes if I lost my virginity before marriage. So I was a good girl and kept my shoes new, so to speak, until I married Cochran.

I was the first black girl to be a proud member of a school club called "The Ladies of King Arthur's Court." It was a club sponsored by very wealthy women from San Francisco. These women would take us on outings—usually to cultural sites like operas and museums. One time, though, they took us to a beachfront amusement park with rides and arcades. After I went on a few rides, I ran into some of my friends who were hanging around a fortune-teller's booth—the kind with dark velvet curtains and a crystal ball. I was a skeptic, predicting, "She'll probably tell all of us the same thing. If you tell enough people the same thing, it's bound to ring true for somebody." Everybody laughed and nodded, but we still all went in to see her.

The fortune-teller took one look at my hand and said, "Oh, dear. I'm really sorry." She hesitated. "Are you sure you want me to tell you?" I was cocky. "Sure," I said. I had the best life, I thought. What could happen to me? She frowned. "I don't know exactly what it is, but I see a double tragedy in your future." I scoffed, "Oh, you tell that to everybody." She shook her wrinkled brow, deadly serious. "No. I don't see it in everybody's hands. But I see it in yours." When I got up to leave, she told me I didn't have to pay her anything because she didn't feel good about what she had said. Well, I didn't feel good about it either. I thanked her for not making me pay. When I went out of the booth and my friends asked what happened, I laughed it off. "Just a lot of garbage." I didn't want to talk about it. I didn't believe her anyway—my life was too good for any kind of tragedy.

From the time I was little, my parents and I always went to the movies. Since I was named after Barbara Stanwyck, I suppose that was predictable. My mother and I were hooked on the love stories. My father would sit through them dying and complaining, shifting his weight from one bun to the other. Occasionally, he'd get his way and drag us to the cowboy movies that he loved. Even at an early age, I thought that watching men running around shooting each other was pretty boring. But I sure went for those love stories.

When I was in the eleventh grade, I saw a movie with my mother called *A Place in the Sun*, with Elizabeth Taylor and Montgomery Clift. My mother had read the book on which it was based, *An American Tragedy*, by Theodore Dreiser. This was the first time I had ever seen Elizabeth Taylor. I was hypnotized—

13

there was something fairy tale–like about her beauty. Eartie and I vowed to see the movie again.

But that was the last time I went to the movies with my mother. The next Friday, Eartie went to get her hair done. When she came home, I noticed her hair didn't look very good. I went over and touched it and it was still wet. She said, "Don't worry, it's all right." That night in bed, she had a stroke.

The next morning, she couldn't speak. Youree and I and Mama Essie rushed her to the hospital. It was crowded. We had to sit in the hallway for what seemed like hours. I sat there holding my mother's hand, while Youree ran around the hospital trying to get someone to help her. She kept trying to speak to me, but couldn't. Finally, they put her in a room. Her whole right side was paralyzed. Later, she tried to get out of bed to go to the bathroom, but her body failed her. She fell back down on the bed. It was agony seeing my beautiful young mother like this.

On Sunday, Uncle Riley came to visit her. He and Youree talked to the doctor. They didn't want to tell me what the doctor said, but I think he told them that if she lived, she would never walk or talk again. I couldn't believe any of it. She was thirty-five years old.

I went to school that week, but my stomach was churning, my hands were clammy, and I could hardly concentrate. I visited her after school each day. There was no change. On Friday, Youree called the school to tell me I should come home. I knew that meant that my mother had died.

I sat on the train on the way home, crying. I realized that I had never cried about anything before, except for the time they wouldn't let me give blood. What a stupid thing that was to cry over, I thought. I couldn't believe how much my life had changed in only one week.

When I got home, Mama Essie was waiting for me. She sat there talking to me, trying to comfort me. I listened, thinking, this isn't real. It can't be real. Never. It is too unbelievable. My mother, my best friend, was gone.

There were two funerals, one in Oakland and one in Grand Cane, where she was buried. At the funeral in Oakland, I couldn't stop crying. I thought my insides were going to come out through my tear ducts. I must have been a pitiful sight, for someone decided that I didn't need to go to Louisiana. One funeral was enough for me. I stayed back home with Mama Essie. I wondered how I would ever learn to stand the pain. There was no one to ask how to do it. I didn't know another soul my age whose mother had died.

Mama Essie and Youree tried hard to keep me from grieving my young life away. They constantly reassured me that everything was going to be all right. They worried whenever I tried to spend a minute alone. But sometimes, all I wanted was to be in my room alone, and cry.

My friends acted awkward around me. They didn't know what to say. And I never talked to them about my mother's death. I guess I didn't think they would be able to handle it. Sometimes my girlfriends would come over and we'd do homework together. The minute they left, Mama Essie would try to pick up the slack. She'd ask, "What do you want to do?" We'd watch TV, go out for a walk, or listen to music. I think I must have locked up a tremendous amount of grief inside of me because I didn't mourn my mother as much as I should have then. It was thirty-five years before I could talk about her death without crying.

Life went on. Youree continued to do his popular radio shows. I continued to go to school. When Youree and I first tried to reach out to each other, it was hard. We were both too grief-stricken. We depended on Mama Essie for everything.

I thought about that fortune-teller—about the double tragedy she predicted. I was in the habit, given my religious upbringing, of talking to God. I told God to have some mercy. I had had enough of tragedy. In the Baptist Church, they always tell you

15

that God won't give you more than you can handle. I wanted to be sure that God knew that I had already had all I could handle.

The summer after my mother died, Auntie Annette wanted me to come down to L.A. to stay for a month. Mama Essie said, "Oh, yes, darling. You need to get away." Good old Mama Essie needed some time off too. With me out of the way, she could really go wild—hitting the racetrack every day.

In L.A., Auntie didn't give me the time to stew very much either. We shopped. We went to lunch. We went visiting. We went to the beach in Venice. At night I would still think about my mother and ask God to take care of her, but I didn't do it with as much sadness and longing as I did in Oakland. Auntie and the L.A. sunshine helped restore my spirits.

My father was glad to see me when I got back. That year Youree and I grew even closer than we had been. He seemed to know that we needed time alone together. If I didn't go to the station with him after school, he'd come home between his two shows and we'd go get an ice cream or take a walk. I started to feel that I was just a girl who would be raised by her father. I would be okay. Youree would take care of me.

I was now sixteen and in my last semester in high school—I was going to graduate in January. That fall, Youree went into the hospital. He awoke one morning and his ankles were swollen. He called Mama Essie and me in to look at them. I put my fingers on them and they made an indentation. His ankles felt all tight. He said he would make an appointment with the doctor and we told him he better—there had to be something wrong.

Youree went to work that day, got busy, and forgot to call the doctor. The next morning his ankles had swelled up even more. They were beginning to hurt. He went to the doctor and the doctor put him in the hospital that day.

Youree told me that he had a kidney ailment. It was nothing to worry about. He just needed to be treated in the hospital for a while. Youree hated the Oakland hospital where Eartie had died and refused to go there. He went instead to a hospital in San Francisco. I talked to him on the phone every day that week but couldn't make it across the bay during visiting hours until Saturday. By then, Youree looked fine. The swelling in his ankles had gone down and they were treating him for his kidney problem.

Weeks passed. A couple of months went by. They still kept Youree in the hospital. Mama Essie and I settled into a routine of visiting him on Saturdays and Sundays and talking to him on the phone during the week. I was confident that he would be better soon—and sure he would be home well before Christmas.

On Veteran's Day, November 11, I had the day off from school, so we decided to go over to the hospital. That morning Mama Essie started cooking up a batch of salmon croquettes. I asked, "What are you doing? We can get something to eat in the hospital or somewhere along the way." She said, "No, I'm taking these to Youree." She had been doing that all along. He would ask her to fry up chicken or fish for him and she would. The doctors had him on a salt-free, bland hospital diet, and he couldn't stand it. I didn't think it was right for him to be eating all this fried food, but I couldn't tell my father and his mother what to do.

In the hospital that day, Youree didn't seem to want us to go. He sat up for a long time and told stories. I was a little concerned when I noticed that his ankles were swollen again. Mama Essie told him to be sure to ask the doctor about them. We left the hospital about 5 p.m. As we walked into the house that evening, the phone was ringing. I thought it was one of my girlfriends. But it was the doctor. My father was dead.

I listened, in a fog, as the doctor said, "He hung on until you left and then he died." I couldn't even cry. It was too much of a shock.

A few days later when the doctor brought over the death certificate, he told me that he was with Youree when he died. Youree said to him that he didn't want to die—he had a daughter he needed to raise. The doctor tried to lift Youree's spirits, telling

17

him that the medicine just needed more time to work. "But it was all too much for him," the doctor told me. "He was still grieving over the loss of his childhood sweetheart. He didn't have the strength to fight the disease."

I was angry at the doctor for not telling me that he might die. But mostly, I was angry at God.

Chapter 2

L ater on, to put me down, or to remind me of how much he had done for me, John would often say that anybody whose parents were dead before they were seventeen was lucky not to be sleeping in a gutter someplace. Well, maybe that's where I might have ended up after Youree died had I come from a less-loving or less-supportive family. But it surely wasn't John who saved me from the streets. It was Auntie Annette and Mama Essie.

The night my father died, I could feel myself slipping into a dark shell. I wanted and needed to talk about all that had happened, but I couldn't. I was numb. How does a teenager deal with the reality that both of her parents are gone? There were things I needed to tell them, and so many more things to ask them. I felt like I wanted to cry forever. And I wanted to scream and scream out—it's too awful, it's too unfair. But instead, I held it all in.

Mama Essie insisted that I go to school the next day and every day. She said that was what my parents wanted for me more than anything else. I must do it for them. So I carried myself off to school. I did what was expected of me.

Auntie flew up from L.A. right away. She helped make funeral arrangements and helped me take care of all the other business I was suddenly faced with. We saw lawyers and settled my father's accounts. Somehow it penetrated my clouded brain that my parents had been saving since practically the day I was born and there was enough money to get me through college. Financially, at least, I would be all right. Their dream for me had not died with them.

At my father's funeral, in a church in San Francisco, I still couldn't cry. I just sat there stone-faced. There wasn't a moment when either Auntie or Mama Essie didn't have an eye on me. When I saw their worried glances, it made me feel as though I had just been released from a mental institution and they were waiting for me to leap over pews and start speaking in tongues. I tried to tell them that I was okay, that I just wanted to be left alone. They must have been thinking, "This poor child. She's lost both her parents in such a short time. What must she feel?" But I wasn't feeling. I was just existing moment to moment, anesthetized, drifting along in a daze.

For all my pain, I had been brought up to be supremely confident in myself, and I remember looking up and lecturing God with the impertinence of a badly hurt but still sassy teenager: "Do you really know what you're doing, Lord? You've taken both of them. And I don't even have any brothers or sisters. If you need some help, God, let me know. I'll be glad to give you some better instructions."

Before Auntie left to return home, she told me that she wanted me to move to L.A. when I graduated from high school. She thought it best that I live with her and go to UCLA. I had already been accepted at UC Berkeley, but she promised to get everything transferred to UCLA. I didn't argue about anything. Whatever teenage desire for independence I once had was now gone. For the first time in my life, I just wanted someone else to tell me what to do, to manage my life for me.

Today we would say that I was in denial, but perhaps that was where I needed to be just then. Every day I woke up in the same fog. My mind and the Oakland weather were in perfect harmony at last.

I endured a rude awakening that Christmas vacation, when the numbness went away. I remember staring at the Christmas tree that Mama Essie had put up and so carefully decorated, trying to make it seem like this was just an ordinary Christmas. I suddenly had flashbacks to Christmases past, when Eartie and Youree, both laughing and carrying on, and making me laugh till my sides hurt, turned the holiday into a celebration of family love. That's when my tears started to flow in torrents again, as they had after my mother's death.

I went back to school in January and came home and cried every day. When I turned seventeen on January 14, I cried. At the rehearsals for my high school graduation ceremony, nobody wanted to march behind or in front of me, because I cried the whole time I walked up the aisle. Here was this grand moment about to happen, a moment my parents had planned for and looked forward to, their graduation as much as my own, and all I could think of was that they weren't there to share it with me.

The tears overwhelmed me. I thought about how my mother would have told me to pray for more strength. So I did. I felt that God—whatever God had in store for me—was all I had to cling to now anyway. I realized that it was time to stop being so angry. I had been taught that God always had a plan, even though it was often unintelligible to us, and that if we had faith, everything would come out all right in the end. Just before my graduation day, I began to pick up the pieces of my young life and look forward to what the future might hold for me.

My parents may not have seen my high school graduation, but half of Los Angeles did. Auntie came up from L.A. for my graduation with an entourage of about twenty-five cousins and relatives, some of whom I had never seen or heard of in my life. It was a beautiful thing. She wanted me to feel that people cared about me—that I had family. I joked with her later, "Where did you get all these people? Did you put an ad in the paper or did you just stand on the street corner and recruit whoever came by?" She laughed at me. "These are your family, Barbara." I loved her for that. Auntie had a way of knowing just what a person needed.

Two days after I graduated, I left for Los Angeles. Mama Essie put up a little resistance at first, but she was no match for Auntie. It turned out that I wouldn't be able to enter UCLA until the following fall because they hadn't received my transcripts in time. Mama Essie used that opening to suggest that I take a trip around the world—to take my mind off things. "She's only just turned seventeen, the child. And you're going to make her go to college? Let her travel for a while." Auntie narrowed her eyes. "And just who would travel with her on a trip around the world?" Mama Essie quickly volunteered to make herself available. But Auntie wouldn't hear of it. "She's not taking a trip around any world. She's coming to Los Angeles and she's going to college. That's what her parents would have wanted."

I packed my things. I consoled Mama Essie, telling her I'd be back in the summer and maybe we could take a nice long trip then. I brought only my clothes, some books, some records, and a few photos of my parents, as if I were going off to a year at school rather than moving out.

When we arrived in L.A., Auntie already had a bedroom ready and waiting for me. She lived in a house in Arlington with Ben, her husband at the time, and my cousin, Doris. It was a big two-story place with nice big rooms. My room even had a telephone in it, with instructions on it: "Don't stay on the phone for more than thirty minutes." I laughed when I saw the admonition. "Who would I talk to, Auntie? I don't know a soul here except you and Doris and Ben."

Since I couldn't begin UCLA right away, Auntie had arranged for me to attend City College, a local community college, for a semester. On the first day, in my English class, I noticed a cute black girl sitting in the front of the class. There were other black students in my classes, but she caught my attention because she was as short as I was. The next day, we spoke. Josie Dotson and I became fast friends. She touched my life from the moment I met her. Much later, when I read an article about John in the *Los*

Angeles Times in which he was quoted as saying, "So much of what I am today, I owe to that woman [me]," I thought, thank God, not much that I am today do I owe to that man. But the influence of Josie—this cute, confident little girl I met that day in my English class—I will feel forever.

Josie had already been at City College for a semester, and she helped me learn the ropes. We started having lunch together at George's, a place across the street from campus that had the greasiest hamburgers I'd ever tasted. Josie's family had the same kind of hopes and aspirations for her that mine had for me. She was from Bracketville, Texas, where her mother and grandmother owned a cafe. Her mother wanted her to come out to California to go to college because she thought Josie would have a better life than if she stayed in the South.

I could talk to Josie about everything, even my feelings about losing my parents. She'd come over to Auntie's house to study with me, and it wasn't long before we'd put our books down and just talk and talk. Josie was a year younger, but she was more mature than I was, I thought. She was very perceptive about people. All through life I would go to Josie for advice and, unfortunately, when it came to John, I didn't always take it.

Josie wanted to be an actress, but I thought she should have been a comedian. She had the knack for coming up with a line at just the right moment that would send everyone into fits of hilarity. Often, I was her straight woman, unwittingly coming out with some innocent, dumb remark that would set up a funny comeback on her part. Typical of Josie was the time when a group of us were at a restaurant and a friend asked her to share a chocolate dessert. She shot back, "I don't like anything chocolate, honey." Then, with a roll of her eyes, "Except my men." That was Josie.

My other close friend was my cousin, Doris. Doris was wonderful to me. She wasn't at all jealous or envious of the way Auntie took me in and treated me just like another daughter. Even though she was older than I, Doris was in her last semester in high school when I came to live at Auntie's. She introduced me to many of her high school friends and soon I was making up for everything in the high school social scene that I had missed earlier. One boy invited me to the senior prom and I went. I was stunned. We danced all night and didn't get home until the wee hours of the morning. If my parents were alive, I thought, I never would have been allowed to do things like this. Maybe by the time I reached twenty-five, if I was still living with them, they would have extended my curfew to 11:30. But Auntie was so lenient. When we asked to go out, she'd say, "Okay. Just let me know where you guys are going and what time you'll be in."

Auntie wanted Doris to go to college and she wanted me to help talk her into it. So I tried to spark Doris's interest by telling her how wonderful college life was. She'd moan in response, "I'm not like you, Barbara. I'm having enough trouble trying to get through high school." I tried another tack: "But what are you going to do if you don't go to college?" She said, "I'm going to get married." I was astonished. "Are you kidding?" But she wasn't.

Josie and I had already discussed this issue of when to marry, even though we had no particular prospects in mind at the time. We knew we wanted to work and do things after we graduated from college. A husband at too early an age, we reasoned, would just be a drag on our lives. Doris, on the other hand, saw things differently. She had met a man named Hillary, the brother of Auntie's best friend. He was twenty-eight, ten years older than Doris. Auntie was devastated when Doris told her she was getting married. She begged Doris to at least try college. She might find herself taking a liking to it. Doris, however, could be just as forceful as her mother. "No, mother, I want to get married."

The uproar over Doris's decision to get married was pretty all-consuming for a while. Auntie went over with us how she had married young herself and the mistake it had been. She ended up divorcing Doris's father and she didn't want the same thing to happen to her daughter. Partly to take up Auntie's side and partly because I truly couldn't understand why Doris would not want to do more in life before she settled down to married life, I kept the pressure on too. "Don't you want to get out there a little bit, Doris? Maybe if you don't like schoolwork that's too serious you could go to a business school and learn to type and file and get a nice job in an office." She brushed me off. "I know exactly what I'm doing." "But why?" I asked, sincerely curious. She paused and looked at me: "I think I love him. And I think he loves me." I was aghast. "But he's so old!" At seventeen, twenty-eight seemed ancient, too old surely to be a romantic lover.

One time I thought I had Doris. She admitted that, in the back of her mind, she once thought she would like to be a nurse. "So why don't you go to nursing school? You'll have a respected profession, you'll be helping people, and you might fall in love with a young doctor." I hoped this last point might be the ticket. But Doris shook her head. "No. I just want to marry Hillary, and I want to marry him now. I want to have a baby. Let me do what I want to do."

So finally, we did. I quit nagging her about school, and Auntie started to prepare for this big wedding with millions of bridesmaids (of which I was one), long flowing dresses, and endless rehearsals.

I had planned to go up to Oakland that summer and help Mama Essie go through all the things in my parents' house, to decide what I wanted, but I didn't go. I was too involved in my life in L.A. and in the wedding preparations. I told Mama Essie that I'd come another time—soon. But there was a part of me, I suppose, that didn't want to go back to Oakland then, even for a visit. I was afraid of seeing that house, with all its memories, all those vivid reminders of what I had lost. And closing up that part of my life by sorting through old things to find what to save as keepsakes and what to chuck out was just too much for me.

The awful word *orphan*—whenever I heard it or thought of it—

brought chills to my spine. It was scary. What had made being an orphan bearable for me that first year was this new life I had stepped into—this whole huge new world that Auntie had opened up for me in Los Angeles. Everything in my life was new—a new school, new friends, a new church, new weather. I thanked God for the weather. At last it was bright and sunny every day, even if the smog did make my eyes water.

I was feeling okay now—but it was still a fragile feeling. Even in my new world, there were times when I couldn't stop the tears from coming. When I would get an "A" in a class, I'd think about how proud Eartie and Youree would have been. I'd go into my room and put on the record player and cry. I knew enough to put on music because it would help keep me out of that low, grieving place I didn't want to go into too deeply. I particularly loved Smokey Robinson and the Miracles. Listening to Smokey sing, "I can't give up hope just because I'm at the end of my rope," would usually stop the tears. Like Smokey, I didn't want to give up this hope I felt in my new life.

After Doris's wedding, she left to live with her husband, and it was just Auntie, Ben, and me in the house. For the first time since I'd moved there, I felt a sadness throughout the house. Auntie missed Doris and so did I. A part of me also felt sorry for Doris— sorry that she wasn't taking more chances in her life, that she had settled for so little so quickly. Although Doris seemed to be very happy, I worried that she was hiding in Hillary's bosom.

When it was time for classes to start at UCLA, however, I was the one who wanted to crawl back into the womb. It's easy to think about how other people aren't taking risks in their lives, but when it comes to taking them yourself, it isn't always so simple. I had gotten used to City College in the one semester I had attended. It was manageable, familiar, and Josie was there. Now I was supposed to enter this huge university and be just one of 20,000 students, all full of ambition, all with their competitive edges well honed. It was scary. I told Auntie that maybe I should stay at City College for a while. But she wouldn't hear of it. "Oh, no. Uh, uh. You aren't going to waste any more time there. You're going to UCLA. That's what Youree and Eartie would have wanted and that's what you're going to do. You'll be all right." As always, I obeyed her. Auntie had been my guardian angel when I had been at my lowest point, and now she became the mother bird pointing me in the right direction, gently nudging me out of the nest, forcing me to try my wings.

As usual, Auntie was right. I was fine. In fact, I loved UCLA. It was beautiful. It was exciting. There weren't very many black students at that time, but we all congregated together and got to know each other. I met people there who became lifelong friends, as well as people—like Senator Diane Watson and former city council member Robert Farrell—who later became leaders in the community.

It was the late 1950s then, and the early stirrings of the civil rights movement were beginning in the South. The black students on the UCLA campus were electrified when the 1956 Montgomery, Alabama, bus boycott attracted nationwide attention. Many of our families had migrated from the South and we felt a closeness to the movement. While we had experienced the more subtle forms of racism and discrimination in Los Angeles, the blatant practices of the South were missing here. So the protests in the South made us appreciate the opportunities that were available to us—opportunities that were being denied to our southern brothers and sisters. It made us want to succeed. Being an exemplary black student was a way we could make a contribution to improving things for black people. We wanted to succeed. Our parents' generation had faced obstacles that we no longer faced. By doing well in a world that had not been open to them, we could make them proud, make their own struggles worthwhile.

During this time, I had to declare a major and make a decision about what I was going to do in life. My dream of becoming the first black female disc jockey had died with Youree. Without him to guide me, without his contacts in the radio world, it didn't seem like a goal I could possibly reach. But my mother's dream for me didn't die. I realized that all through school I never once had a black teacher. At UCLA as well, none of my teacher's were black. Becoming a teacher, I felt, would not only be a tribute to my mother's unfulfilled dream for herself but would be a way that I could do my part for the civil rights movement. As a teacher, I could be a role model for young black children and provide inspiration that might help them make the most of the new opportunities the civil rights movement might open up for them.

I was a serious, focused student, but I also had my share of good times at UCLA. A semester after I got there, Josie transferred to UCLA. I didn't see too much of her at school because she was a theater arts major and thus was always on the north side of campus, trying out for plays or backstage building sets. But, like Auntie, Josie always pushed me on to new heights. There was a black fraternity at UCLA, Kappa Alpha Psi, and for their fraternity basketball games they needed cheerleaders. I had received an invitation to try out for the Kappa cheerleading squad and I made the mistake of mentioning it to Josie. She said, "Oh, I'm definitely going to try out. I was a cheerleader in high school." I said, "Good for you. You do it. I'm not going to." She frowned at me and I knew I wouldn't hear the end of this soon. "Oh, Barbara, come on. If they've invited you to try out, somebody thinks you're cute. It will be fun." She kept working on me and I kept saying no. But when it came time for the tryouts, I was there with Josie.

I looked around me and was overwhelmed. Hundreds of girls were trying out to be Kappa cheerleaders—and they only wanted five. I told Josie it was a waste of time. I wanted to leave. She gave me that "Don't quit on me" frown again, so I stayed. When it was my turn, I got up and sort of danced and pranced around stupidly, trying to do what Josie had done. They picked a tall girl for the middle, two medium-sized girls for next to her, and for the ends they wanted two short, tiny girls. I just about fell over when they read my name along with Josie's.

After several weeks of rehearsal, there I was at the Kappa fraternity basketball game, ready to run out and make a fool of myself in front of a huge crowd that only wanted our side to throw the big round ball through the small hoop more often than the other side. It was a really big night for the fraternity. Every black student, not just from UCLA, but Kappa members from other colleges in the area, had come out for the game. I remember meeting a man named Tom Bradley that night, a Kappa alumnus who was back to see the game—he later would become mayor of Los Angeles. When it was our big moment, I ran to center court with Josie and the three other girls. We shook our booties and pranced around and were just as ridiculous as we could be. But Josie was right—it was fun.

Next Josie got it into her head that we should try out for the UCLA cheerleader team. "It's about time they had some black cheerleaders," she reasoned. I'm sure she was right about that, but I'm not sure Josie and I, as tiny as we both were, were the best candidates. Unlike the Kappa cheerleaders, the UCLA cheerleaders had to jump around with these huge pom-poms. I didn't think I could even lift them, but Josie dragged me to the tryouts. This time, fate was kind—we weren't chosen.

I didn't have a car when I was going to school so I hitched a ride with a variety of people. A girl named Marlene Edwards lived right around the corner from me and several of us rode to school with her, each paying her a quarter a day to cover gas. After Marlene graduated, I rode with a guy named Herbie Avery, who later became a doctor. Herbie was from Georgia, and he hated it when we would call him "Georgia Boy."

One semester, I had to take zoology, and one horrible day that semester, after spending hours in zoo lab cutting up a frog, I left class and realized Herbie had already left. Usually, I loved walking across that beautiful campus, with its sloping green lawns, tall trees, and old brick buildings. But this particular afternoon, though the campus may have been just as tranquil, I wasn't. I felt tired, stranded, and was irritated at the prospect of having to take the hot, meandering bus ride home. I asked around among the black students still on campus to see if anyone was driving my way. I was relieved when a friend, a guy named Mike Grubbs, told me that there was a fellow still on campus who would probably be able to give me a ride home. I sat down and waited with Mike for this fellow to come along.

Little did I know then that I was about to embark on the longest, bumpiest ride of my life. Had I known, I would have gladly taken the bus.

Chapter 3

"John Cochran, this is Barbara Berry," Mike said when my potential ride came along. John had a nice, pleasant face, with bright eyes and a quick smile. "I know who she is," he said. I wondered how, but simply greeted him, "Hi, John." Clearly the friendly type, he continued, "How are you, Barbara?" That was my opening: "Fine, but I have a slight problem." "What's the matter?" he asked. "I need a ride home," I almost pleaded. "I heard you were going my way. Do you have room for one more?" He told me he did and would be glad to take me. I sighed with relief and grabbed my books.

We started walking toward his car, making the usual conversation that students make—about school, classes, or students we both knew. John mentioned that he was a member of the Kappa fraternity. I figured that was how he knew me. Like hundreds of other poor souls, he had probably witnessed the spectacle I had made of myself, prancing around and shouting cheers on the basketball court at fraternity games. I didn't mention it. As his other riders joined us in our walk to the car, John told me that he knew my friend, Josie. She often rode with him to school. Maybe that was how he knew me, I hoped.

When we got to John's car, one of the other girls said she wanted to stop and get something to eat at Thrifty Drugs in Westwood. John agreed to stop. While we were in the drugstore, he politely asked me if I wanted something. I told him I wasn't hungry, which was true. Cutting up that frog in zoo lab had ruined my appetite, but I also didn't want to embarrass him, in case he might feel obliged to treat me and didn't have much money.

After Thrifty's, John dropped off the others and asked if I would mind if he stopped at home before he dropped me off. He had to pick up something and then go to his job at the post office. My house was on the way there. I didn't mind. I figured I would still be standing at the bus stop if he hadn't come along.

John lived with his parents in a humble but well-maintained house on West Twenty-eighth Street. I went into the house with him and met his mother, Hattie. I liked her instantly. She was warm, sweet, and gracious. She talked to me while I waited for John. Clearly very proud of her family, she told me about John's two sisters and his younger brother. Not only was John in college, but so was his sister. As Hattie talked, I heard the same kind of homespun goodness and optimism that my own mother had had. What a nice family, I remember thinking.

John dropped me off at Auntie's. I thanked him and we both said we'd see each other around campus. Over the next few days I didn't think much about him or the ride home. He just seemed like a friendly, hardworking student from a nice family.

A few days later, when I saw Josie, I mentioned to her that John had rescued me from an agonizing bus ride. I knew that she rode to UCLA every day with John, his sister, and his girlfriend. I was surprised when she indicated that she didn't think he was as nice as I had found him and asked her why. She described a disturbing incident that had happened about a week ago on the way to school. As they were waiting in John's car at a stoplight in a white neighborhood one morning, a black woman got off a bus and made her way to a nearby house, apparently to do housework. John and the others in the car saw the woman as an absurd figure and made fun of her, putting her down for being "just a maid." It angered Josie and she defended the woman, telling the other girls and John that the woman was working hard—probably at the only job she could get—and that she deserved their respect.

32

Josie was still fuming about the incident. Josie and I both had grandmothers who had done domestic work to help support their families. It was one of the few ways for black women of their generation to make an honest living. Josie's grandmother had saved her money from her long and hard years cleaning other people's houses and was eventually able to buy a little cafe in the small Texas town that Josie was from. It was the money her grandmother and mother earned from that cafe that helped send Josie to school, but it was the domestic work that had given her the stake she needed to go into her own business.

Josie thought John and his friends were immature, to say the least. I agreed, but made an excuse for them. "They're a little younger than you are, Josie." At that age, a couple of years seems like a lifetime. "They'll wise up in time," I assured her. In the weeks that followed, I ran into John around the campus occasionally, and we would wave or talk for a minute, but I still didn't think that much about him. I was focused on finishing my last semester on campus. Next semester I would be doing my student teaching—the last step before I got my degree and hit the real world.

I dated several guys during my years at UCLA. I remember in particular a Kappa member who had transferred from Ohio and also a few others. Dating then, however, was nothing like it is today. This was years before the sexual revolution. The mores of those times—the fifties—were very different from what they would become for the students of the sixties and seventies. There were very few couples sleeping together—at least few that I knew of. Dating was about going out and having a good time. People became friends first, long before they got serious.

So there was no one that special in my life when Josie came over one afternoon and said, "John Cochran told me that he's interested in going out with you. He said you seem like a lot of fun. I told him you were the dullest thing I had ever met." I laughed at Josie and asked her, "What happened to his girlfriend?" Josie shrugged. "She got wise, I guess. You don't want to go out with him, do you?" I told her that I wouldn't mind. He seemed nice enough. She rolled her eyes, but then must have carried or sent the word back to John that I would accept an invitation. It wasn't long before I ran into him, and he asked me if I wanted to go to a party with him.

And so our first date was a fraternity party. As I suspected we would, we had fun. We both liked to laugh, and we knew a lot of the same people who frequented those fraternity parties. We danced the slow dances together—John didn't dance fast, although his friends would drag him out to do the one fast dance he knew, the "turkey," a particularly goofy dance. We dated off and on that semester, both of us still seeing other people, and generally had an easy, comfortable time together.

As stable as things had been for me during those years at UCLA, I knew from the experiences I'd had in my young life that change could come at any moment. Still, it always took me by surprise. One day, toward the end of that semester, I came home from school and Auntie told me that her husband, Ben, had left and wouldn't be coming back. I was stunned. I had been wrapped up in myself and hadn't noticed anything wrong. Auntie said that it was just a gradual thing. They had fallen out of love. Ben had asked Auntie to say good-bye to me. I thought that there must have been more to it than met my eye, but I honestly didn't know what—and Auntie never mentioned Ben again.

Auntie wanted to go to El Paso—where she had some close relatives—to get away for a while. I told her I didn't mind staying in the house alone. I would be student teaching in the fall and I'd be fine. She shook her head. She had already spoken to an old family friend, Patience, who said she would be happy to have me live with her.

Patience had been my mother's best friend when I was growing up in Oakland. She had a daughter—also named Barbara, of course—and she and I had often played with one another over the

34

years when our mothers were close friends. After my mother died, Patience and her family moved down to L.A., but she kept in touch. When it was time for my high school graduation, Patience made a special trip up to Oakland to help me shop for a dress. Even though Mama Essie was fairly hip, she was still a grandmother and Patience didn't want to leave that task to her. After I moved to L.A., Patience tried to look out for me whenever she could. She took me to her church—Second Baptist—and I joined it. We both loved to shop, and she would often take me shopping on Saturdays. I'd help Patience pick out things to send to the other Barbara, who was attending a small black college in Texas.

That summer, Auntie helped me move into Patience's house, closed up her own house, and left for El Paso. I missed Auntie, but I was content living with Patience. In the fall, I began student teaching. Patience dropped me off at the elementary school every morning and then went to the post office, where she worked in the nurse's office.

The parties I went to on the weekends that semester were my only connection with the students and the campus, since I was spending every day student teaching in an elementary school. I continued to date John, as well as other guys, and as far as I knew John was seeing a couple of other girls. At one party that John and I had gone to together, a girl kept playing footsie with him under the table. I can't say that he did anything in particular to encourage her—but he seemed to enjoy having one girl with him and another one after him. I didn't think much about that at the time—after all, we weren't going steady or anything.

The first time John ever got a little peeved at me was the night of a dreadful rain. He came by to pick me up to go to a party just after it started to pour. I can't stand heavy rain and told him I wouldn't be able to go out. I didn't like getting wet or driving in a rainstorm. He looked at me, in a state of shock and disbelief. To me it was just another fraternity party—what did it matter if I missed one of them—but he was quite disappointed. He didn't say so then—he was still courting—but I wouldn't be surprised if that was the first time he thought that I wasn't all there. I told him to just go on alone if he wanted—it would be all right with me—because I really wouldn't enjoy myself in that downpour. He decided to stay, and we just watched TV. But that wasn't the end of it. For the next few decades, John would remind me of that night every time it rained.

My final semester, spent student teaching, passed in a flash. I graduated from UCLA on a January Thursday and on the following Monday I was to start my first teaching assignment. Auntie came back from El Paso for my graduation. She gave me her car because she had decided to sell everything in L.A. and make a permanent move to El Paso, where she planned to open a beauty parlor. Auntie rounded up all of my long-lost relatives, and she and Patience threw me a graduation party. Those strange relatives all came and so did Josie and a couple of other close friends. When I got home that night I was pleasantly surprised to see that John had dropped off a beautiful bouquet of flowers for me.

After Auntie left for El Paso on Sunday afternoon, Patience and I sat down at her little kitchen table. I thought she was tired and would probably take a nap, leaving me a little time alone to think about what I would do the next day—my first day as a teacher. But she didn't take a nap. She sat there and said what I was thinking—how proud my mother would have been on this, my second graduation day since her death. If only she could have seen me graduate. I started to cry, but it was a good cry. Patience and I sat there for hours talking about my mother, remembering the fun we had had together. It was our way of bringing her back to us, of making her a part of so important a day in my life.

The next day, I began to fulfill my mother's dream. I drove to the Wadsworth Avenue Elementary School, in the heart of the inner city. It was 97 percent black. The only white children there were children who had one parent who was black. I was filled with excitement and enthusiasm as I looked out at all those young faces. Once again, I was entering a new world, as I had when I entered UCLA. But this time, I didn't have the luxury to agonize about it. The children immediately demanded all of my attention.

My classroom was sandwiched between the classrooms of two seasoned middle-aged teachers. Both were black and both were from New Orleans. These were good-looking sisters. They had taught school since they were twenty-one. Both were married but had no children. They spent their money on themselves and wore expensive silk suits and attractive alligator shoes. Next to them, I felt like Rebecca from Sunnybrook farm in my modest little Lanz dresses and simple shoes. Finally I had some black teachers of my own, for these two women took me under their wing and taught me more about teaching in six months than I had learned in four years at UCLA. And, they turned out to be the best role models I could have ever had—in my first evaluation, my principal rated me outstanding.

Now that I was a career woman, Josie thought that it was time I experienced what it's like to live on my own without adult supervision. She and another friend, Yvonne, were getting an apartment together and needed a third roommate to make the rent affordable. I figured that after all those Sundays at church in Oakland—and here in L.A. at the Second Baptist Church—my morals were good enough. I had kept my "shoes new" all this time, the way my mother said to, and could probably be trusted not to go crazy on my own.

We first rented a modern upstairs apartment on Second Avenue in a little turquoise and white building. Josie and I shared a bedroom, but they gave me the extra closet because, after all my therapeutic shopping trips, I had tons of clothes. I didn't have to buy any furniture at the time because Josie and Yvonne had everything, but my contribution was a wall-to-wall Magnavox stereo. It was my first big purchase in life—and Smokey Robinson sounded sensational through those stereo speakers.

Although we all liked the apartment, it was small. One of us heard about a house that we could rent for next to nothing, because it was going to be torn down in a year to build a freeway. So we moved. Now we each had our own room and there was a huge kitchen and a huge living room to share.

We had great times in that big house. Back when Josie and I first met, as Doris was rushing off to get married, we had fantasized together about a time like this after college. Now here we were—Josie, Yvonne, and me, all career women with interesting jobs. We earned our own money. We spent time together and we spent time with the guys we were dating. We had great parties. And we didn't have husbands to account to or children to take care of. It was a wonderful time of life.

Josie was working then at Inner City, a cultural center that put on plays and dances and sponsored art shows. She was also in a few plays around town—when there was a black production or a play with a black character. Often, Josie's theater friends came over to the house. I loved them. They were sophisticated, interesting, and worldly—wiser than most of the other people I knew at the time.

At some point I mentioned to Josie that even though I loved the life we were leading, eventually I wanted to get married and settle down. She looked at me with mild alarm and amusement. "You better be careful what you ask for, girl. You just might get it."

I don't remember exactly when it was that I stopped seeing other guys and started seeing John every week. We just sort of slipped into it. I recently asked one of my friends from that time if she could remember when it changed or what had happened to change things, but she couldn't recall either. "It was just something everybody seemed to know," she told me. "John and Barbara will one day get married."

It was the fall of 1959. John had graduated from UCLA the previous June and was now attending law school. It seemed like all of our friends—all the couples we knew from our UCLA days—were getting married. This was the end of the decade of the fifties, that postwar period of enormous optimism when it seemed the country would soon be filled with happy homesteads—of all races—each with an Ozzie-and-Harriet–like couple and 2.8 children snuggled warmly inside. Marriage was more than in the air among our circle of friends. It was a mark of the times, an important symbol of accomplishment, and John, ever alert to trends, would certainly have sensed it. One time when we were out together—I don't even remember where we were or exactly what day or night it was—he asked me to marry him. Without thinking about it much, I said I would. If it was the time to marry, John was certainly a proper prospect for a husband—reasonably good looking, charmingly well mannered, respected by all for his hard work and intelligence. He must have been pretty sure that the answer was going to be yes because he had a beautiful emerald-cut diamond engagement ring ready for the occasion.

When I showed Josie the ring and told her that John had proposed, she wasn't very pleased. "Are you sure you know what you're doing?" I said that I thought I did. John was a nice guy. He's from the same kind of background as I am—college educated, churchgoing, responsible, from a good family, with attitudes toward marriage and family similar to my own. We had even discovered that we had been born close together, both near Shreveport. He'd probably make a wonderful husband and father. It was the kind of thing I had read about in sociology books. I told her, "A woman should try to marry a man who has the important things in common with her."

"But you could get anybody you want," Josie protested. "You don't have to marry *him*. Tell me, do you even love him?"

She was bugging me the same way I had nagged and harassed Doris when she announced her intention to get married. "Yes," I told Josie. "We love each other, and we really get along well. We have respect and admiration for each other, too, and that's important."

39

She still wasn't satisfied, so I told her that even though it may not be like the couples in the romance novels, who have difficulty relating to mere earthlings, who aren't able to do anything or remember anything because they are so in love, "John and I have something deeper." I pointed out that some of the couples we knew from college who couldn't keep their hands off each other had gotten married and then either separated or divorced after six months.

Eventually, Josie let it go. She had never liked John, but she wanted to communicate that she supported me in my decision to marry him. "Look," I told her, still wanting her complete approval, "there are no guarantees in life, but I think I'm doing the right thing."

Chapter 4

During the time we were engaged, John and I began going to church together. We were both members of the Second Baptist Church, a pillar of black middle- and upper-class life in Los Angeles. Church membership included many professionals—judges, doctors, lawyers, teachers, principals—as well as prominent black politicians and celebrities. John, inspired by the success of many of the people alongside him at Second Baptist, was clearly motivated to work hard and take his place among the important leaders in the congregation. I was also encouraged by the older teachers and principals I met there to be the best that I could be. We were the up-and-coming young couple, the darlings of the congregation.

As I was sitting in church with John one Sunday, I thought about my parents. I looked ahead to my married life with John and imagined us doing the kind of things that Eartie and Youree had done together. I could hear the same kind of fun-filled banter at the dinner table. I could see us going to church together and to school functions for our children, as my parents had. I thought of us sharing ideas, sitting down and working out goals for our life together, and supporting each other in our separate endeavors toward those family goals. I could sense a warm affection and love between us that would grow even stronger over the years. I envisioned raising our children so as to instill in them the values I presumed we shared, the same values my parents had instilled in me—love and respect for each other, a desire for knowledge, a belief in God—combined, naturally, with, a good sense of humor. I took John's arm as we left the church that day, feeling we would have a wonderful marriage. I looked forward to celebrating our twenty-fifth and fiftieth wedding anniversaries together and, of course, all the happy years in between.

It no longer bothered me that my relationship with John wouldn't have made it into any of the Harlequin novels I liked to read. Since I was still able to focus on my job and other things in my life through my engagement, I concluded that I simply wasn't as emotional as the women in the novels, women who would go into an altered state when they were in love, unable to do anything except talk about and feel their love. I was a more rational type of person, I thought, and there was nothing wrong with that.

Shortly after John proposed, we got together with John's parents to tell them about our engagement. John's mother seemed surprised. Perhaps she had always expected him to marry his high-school sweetheart, whom he had been going with when he and I first met. But despite her apparent surprise, Mrs. Cochran accepted her son's choice without apparent reservation. I had always liked John's mother, and my relationship with her became even stronger once John and I decided to marry. She treated me as though I were one of her own children. John's father also seemed to approve of me. I fit his image of the kind of wife he thought John should have at his side—sweet, pretty, educated, churchgoing. John's father, who worked at Golden State Mutual Insurance Company, a black-owned company, had done quite well in life. Having come from the South as my father had, he also managed to build a successful career and give his children opportunities he had never had. He was pleased with his son's drive and ambition, and he knew I would be the kind of wife to encourage and support John in his efforts to become a successful lawyer.

After John and I set the date for our wedding—July 10, 1960— I called Auntie in El Paso. I knew that things had been going well for Auntie since she moved. Not only did she have a thriving beauty parlor, she had met and married a wonderfully sweet man, Raymond Nelson. I liked Raymond, and since Auntie was the closest person I had to a mother, I wanted Raymond to give me away at the wedding.

Auntie could not remember meeting John. When she left L.A., he was just one of several guys who had come by to pick me up for a date. I told Auntie that she would love John. He was a nice guy from a nice family, and he planned to be a lawyer. Best of all, he had a great personality and a tremendous sense of humor. She was delighted that I had found someone so wonderful and said that Raymond would be honored to give me away.

For Auntie, I was fulfilling a dream—my life was perfect. I had graduated from college, I had established myself in a career, and now I was marrying a nice young man who had very good prospects. In Auntie's view, her daughter, Doris, had fared less well. By now, Doris had three young children and was living in Little Rock, Arkansas, with her husband, Hillary. He had left his job in L.A. and relocated his family because he decided he wanted to go back to school. Doris was unhappy about the move and she and Hillary weren't getting along. So, to Auntie, I was the "daughter" whose life was going as it should, as she would have wanted things to be for Doris.

John and I were described as a "popular couple" in the engagement announcement in the *Los Angeles Sentinel*, L.A.'s newspaper for the black community. And all popular couples seemed to have engagement parties, so we decided to have one too. At that time, sophisticated white people in Los Angeles usually held their engagement parties at Chasen's. The middle-class black answer to Chasen's was Tommy Tucker's Play Room. Many times John and I had been to Tommy Tucker's for dinner, the "in" place for blacks then. We had met the owner, Tommy Tucker, and his wife, Gussie. John admired Tommy, who was a very progressive and successful businessman, and Tommy liked John—he claimed to see things in John that reminded him of himself when he was younger. This was where we would have our own engagement party.

We invited about thirty or forty people to Tommy Tucker's on Valentine's Day, 1960, to make our engagement official. Tommy outdid himself, decorating the place with hearts and flowers and red and white streamers in honor of the occasion. At some point in the evening, one of John's friends or family members got up and reminded everyone that they were invited to celebrate John and Barbara's engagement. After that, the champagne started to flow, as our guests toasted us again and again. Then John got up and gave a little speech. Everybody gasped and marveled at his sophisticated vocabulary and syntax. He sounded like a UCLA professor, maybe in an art appreciation class. I guess that John, still a first-year law student, wanted to strut a bit of the stuff that he hoped would one day make him a successful attorney. He went on and on about things like his "unswerving devotion and loyalty" to me. At the end of his speech, I smiled and said thank you. But my thoughts were darker than the sweet words I uttered. "That doesn't sound like the John I know," I thought, troubled by the suspicion that the speech hadn't been a simple expression of his feelings for me, but had been crafted to impress the other people there. I wished he had just spoken to me sincerely, from his heart. The problem—as I was to learn much later—may have been that this was indeed the real John.

Our engagement party was the beginning of a whirlwind of nuptial activities. My life was suddenly filled with lists, invitations, strange events, and thank-you notes. My friend, Myrna, gave me a kitchen shower. Fifty women bearing boxes of pots and pans crammed into her house—my friends from college and my newer friends from teaching. I thought the silly shower games we played at this gathering were pretty inane, but I hadn't seen anything yet. Next, Yvonne and Josie gave me a lingerie shower. At this shameless event, I was bombarded with a steady stream of sexual innuendos and jokes. Yvonne had grown up in New York, and Josie, of course, was part of a sophisticated circle of theater people. They were both so much more worldly than I. Led on by Josie and Yvonne, everyone at that shower teased me relentlessly. The torture was particularly agonizing since they all knew I was still a virgin. I thought I would die. I had always been embarrassed by anything to do with sex—and this was everything to do with sex. I barely survived the ordeal, but probably grew up a little in the process.

I was much more at home shopping than I was hearing what fifty women had to say about sex. Patience helped me shop for my dress, shoes, gloves, and veil and helped me pick out the color scheme for my bridesmaids. We traipsed all around town, making several trips, before I found what I wanted.

Between these numerous shopping trips, I managed to find time here and there to spend with John—the person who was the cause of this flurry of activity. He was excited and was counting the days until our wedding. We'd occasionally see a movie or go to a party, but we spent a lot of our time together dealing with the practical things we needed to do. We looked for a place to live and found a cute duplex in a nice area on a pretty, wide street. John went with me to pick out furniture for our new home, but he deferred to my taste on most of the items. While I fussed with the caterer for the reception, he made the arrangements for the hotel on our wedding night and for the honeymoon. We were both good at this kind of planning and handling of details. Friends complimented us on being such a "take-care-of-business" couple.

John and I, according to everyone we knew, were the absolutely perfect couple—to everyone except Josie, of course. Almost up to the day of the wedding, Josie continued to probe with her piercing questions, like, "Do you really love this guy?" or "Tell me what you feel, Bobbie." In the midst of my busy, hurried wedding preparations I would try to brush her off, "No one is ever 100 percent sure about anything, Josie." I'd tell her I didn't have time to talk—I needed her to help me finish addressing the invitations or whatever task we were in the middle of doing.

It was typical of me then to fill my life with a million things and keep myself so busy that I didn't have time to think about my feelings. I forged ahead without looking back—or inside. Perhaps this was the legacy of having suffered the enormous emotional trauma of losing both parents so close in time. Whatever the reason, here I was about to enter another new phase of my life, and I still did not have real closure on a previous part. I had never once returned to Oakland since I moved to Los Angeles after my father's death. Every year, Mama Essie would ask me, "When are you going to come up here, child? When are you going to go through the things that your mother and daddy left and take what you want?" And every year I would have some new excuse, some legitimate reason why I was too busy to go there. I had seen Mama Essie on her visits to Los Angeles and she was as lively as ever. She had met a man and was getting married herself right around the time I was. Eventually, Mama Essie gave up on me. She finally sold the house that belonged to my parents and moved to another town north of Oakland.

The wedding was going to be held, of course, at Second Baptist, our large, elegant church. In what seemed like no time, July arrived and we began wedding rehearsals. The wedding director was an ageless, timeless woman who it was said had directed five million weddings in the last few years. She had attained such stature in the community that she was called "Madame" Outley, rather than just plan Mrs. or Miss Outley. Like a drill sergeant, Madame Outley quickly whipped us and our motley crew of ushers and bridesmaids into an elegant, coordinated unit, each person ready to perform his or her important part.

When the fateful day arrived, I was calmer than I expected I would be. In my dress, I walked to the entrance of the church. While someone straightened my train, I looked inside and saw that the huge church was filled with people. I caught a glimpse of the bridesmaids and was stunned by how beautiful they looked. At the last rehearsal, they were just an ordinary-looking group of women. Now, standing in their rainbow-colored dresses, they were radiant.

I stood there for what seemed an eternity, waiting for the music to change, our signal to start down the aisle. When it did, Uncle Raymond took my arm. It seems remarkable, looking back, but I don't recall a single tear coming to my eyes any time during that day. I'm always sentimental at weddings—I've never attended another wedding before or since where I didn't cry. Perhaps I was too excited.

Raymond and I started down the aisle, which now seemed half a football field longer than it had been during the rehearsals. As we approached the altar, there was John, waiting, gazing at me. He looked wonderful in his tuxedo. He was smiling and I smiled back. In that moment, maybe for the first time since the death of my father, I felt that my future was safe, for here was a kind and gentle man who could be trusted never to hurt me.

After the ceremony, we posed endlessly for pictures, and then went to the reception. We danced and talked and laughed with our friends and families. Everyone wanted us to leave after we were there only a short while—they were anxious, I suppose, for me to lose my virginity. But we lingered. We were having a great time.

When we did leave and got into the limousine, Yvonne tried to get my attention, gesturing to me to move over closer to John. I was sitting on the other side of the car from him. I didn't notice Yvonne or anything right then, for this was the moment a cold panic chose to strike me. "What have I done?" I thought. "The man in this car with me is my husband. I'm supposed to be one with him for the rest of our lives. Oh Lord, have I made a mistake that will ruin both our lives?" The photographer must have jarred me out of my trance, insisting that I move closer to John for a picture. It was just one more request, in a day of thousands of requests, a day that marked the beginning of my role as the she half of the "model couple" that John and I now were—a life where we would have to meet the expectations of our public regardless of what the reality between us might have been. I did what the photographer asked. I moved closer to John. I smiled. The camera clicked a few more times, adding to the record. And then we drove off.

PART II

Barbara Cochran

Chapter 5

The hustle, bustle, and stress of our huge wedding, added to all the long and difficult preparations leading up to it, had taken their toll on me, and I was running on close to empty. It wasn't until we reached our hotel that John and I were finally able to relax a bit. Every grand wedding is in some ways a parody of itself, and John's fine sense of humor, especially his appreciation for the absurd, broke the ice for me and lightened my spirits. I was elated to find that he too had found funny many of the moments that had nearly broken me up at my own wedding. As we laughed to near choking about some of the events of the day, trying to top each other in the absurdities we recalled, the sense of disquiet I had felt in the limousine evaporated. Under the always-on Mr. Up-and-Coming Cochran, there seemed to be a man who didn't always take himself—or his public pose—too seriously.

Making love our first night was sweet, tender, and passionate. No longer a virgin, I was better able to appreciate what all the fuss at my lingerie shower had been about. The sexuality added a special dimension to my relationship with John, and I looked forward to its being one more nice part of our future together. The man I had married could be kind and gentle and, at least in my presence, down-to-earth. If his public posturing did not always ring true for me, that really wasn't my problem. It may have been part of what he saw as necessary preparation for his life at the bar, and I did not know enough about that life to say when he was or wasn't going too far. My feeling that night was that we had so much to look forward to together; the foreboding voice I had heard in the limo had been quieted by these few hours in a hotel room with my new husband.

If John had already started to pay the price of being a successful lawyer, he was surely not yet reaping its rewards. As a struggling law student he had limited funds, and we had to settle for a honeymoon that wouldn't be too expensive. We picked Santa Catalina Island (which Angelinos generally shorten to just plain Catalina), a short boat ride out of Los Angeles Harbor. We spent five wonderful days on the island, walking on the beach, bicycling, strolling through the quaint shops, and eating at the restaurants. It was one of the most relaxed times we would ever spend together.

After those magic few days, we returned to our little duplex, our first place together. Although we both found it a corny tradition, John carried me over the threshold, both of us giggling through the ancient ritual. John worked at his father's insurance company over the summer break and had to get back to his job the next day. I was off from teaching for the summer, so I tackled the chore of opening boxes and putting away the scores of wedding gifts.

What a beautiful summer that was, with life easy and new. We had good times alone together, but also frequently had friends over to help us break in our new place. When my card-playing friends, Myrna and her husband, George, came over, we tried to teach John how to play cards. He was definitely not a born card shark—he kept saying that he didn't know a heart from a diamond. We all laughed at the line. Many years later, I recalled his words and thought how true a metaphor he had chosen for himself—he never did seem to understand the difference between giving a woman his heart and having a jeweler box up a nice diamond for her. Of course, there are many people who would say that it was my value system that was cockeyed, not his.

September came. I went back to teaching and John returned to

52

law school. We were both very busy, but we still enjoyed our dinners together—joking, laughing, talking about whatever had transpired that day. Even though John's study load was heavy, we found time for an occasional movie, for going out to dinner on Saturday nights, or for seeing friends.

Church, of course, continued to be not only the center of our religious life but as well an important part of our social and political life. And I must admit that it was also a weekly fashion show. We ladies each had our own style and our own standing to uphold, pressures from which the men were not entirely exempt. Although John had a reputation as a tasteful dresser, his wardrobe could hardly have been called conservative. In those days he didn't have the money to pour into clothes that he does now, but he was never a plain navy or gray suit kind of guy. It was part of his personality to dress as distinctively as his sharp sense of good taste and his wallet could afford.

Church was also the place where we got firsthand information on the progress of the civil rights movement. Dr. Martin Luther King, Jr., was a friend of our minister, Dr. Thomas Kilgore, and visited our church often; I remember many occasions when we listened enthralled as he addressed us. After one speech, I met Dr. King at a reception held for him at the church. He was a powerful, intense, moving speaker, but I was struck by the fact that off the podium he had an ordinary homespun quality. Warm, friendly, and funny, he was quite clearly a man who enjoyed the company of people in the humblest of settings. Only a sharp student of human character would have picked him out of a church supper crowd as a man whose unique vision and courage would make him a world leader, someone who would eventually win a Nobel peace prize.

But even at that early stage there could be no doubt that the civil rights movement led by Dr. King was changing America. It had already brought down many old barriers once thought to be indestructible, and would soon play a major part in rewriting the country's Jim Crow laws. But possibly most important for blacks was that it altered forever the image that we had of ourselves and our place in society. No matter how long it would take for some of the social changes to be brought about, it was already true that we would no longer tolerate in our own heads the concept of second-class citizenship. John and I often talked about where we fit in during this period of historic change. We were proud, as young black professionals, to have a part to play in turning Martin Luther King's dream of an America characterized by opportunity for all, and goodwill among all, into a daily reality for America's blacks. What more could a young couple have hoped for! A plan for the future based not on personal acquisitiveness or ambition but on uplifting principles, to be lived out with a loving partner.

During these first few months of my marriage, I was very content. Occasionally—perhaps a visit with Josie sparked it—the old question would enter my mind: "Did I rush into marriage too quickly?" And the answer would always be no, I had done the right thing; everything was going just fine.

It goes without saying that after we were married I discovered in John certain disturbing qualities I had never noticed through the courting period. But this happens in any relationship, I figured. And who knows what dreadful personality flaws he had discovered in me since our wedding. One of the more troubling traits I now noticed in John for the first time was how highly excitable and volatile he could get—often over what seemed to me to be some trivial or imagined slight. Something would happen at school or work and he would come home in a highly agitated state. From dealing with children and their upsets all day, I had learned how to put troubling things in perspective, or to jolly a youngster out of his or her distress. I tried on John some of the methods I used with children. I'd say to him, "You're upset about what *he* said? I can't believe you'd let what such a person said upset you." Sometimes it worked. He'd laugh and say, "You're right. I don't know why I let a guy like that bother me."

Now that I was around John every day, I was amazed at how much nervous energy he had. He would get all wound up and start talking rapidly, whirling his pen, or tapping his foot. He just couldn't seem to relax, always planning, organizing, working, then jumping up and pacing back and forth. Even in church, in the middle of an inspiring sermon, John would have a paper and pen out, adding to or fixing up his list of things to do for the week.

Nor could he sleep. Not just on work days, but on weekends, he jumped out of bed early in the morning. This would not have been a problem for me except that John was the kind of guy who, once he was up, wasn't happy until I was also up. He dropped things and knocked into things until I'd finally stir. Of course, he'd then say, "Did I wake you dear? I'm so sorry." On Saturdays, he'd want to be out of the house early to get to the barbershop for a trim or to have his nails done. On Sundays, he always wanted to go to the 8 a.m. church service. When I told him I preferred that we go to the 11 a.m. service, he'd moan that I was going to sleep my life away. And I'd answer that sleeping in till 9:30 on a Sunday morning was not going to cost me my life, or my soul.

Perhaps the thing that bothered me most about John in those early years, something I also hadn't been aware of before we married, was his tendency to gossip. I had the standard preconceived notions of the time about the differences between men and women on this score. I accepted the stereotype that women gossiped, but men didn't. John proved me wrong. I would often hear him on the telephone gabbing with his men friends—talking about other people in our circle of friends, people we liked and socialized with, people John treated to a full dose of his resplendent charm at every meeting. Once he referred to a friend who didn't straighten her hair as "too proud to press." That kind of remark would have been seen as catty coming from a woman, but here it was coming from my husband. Thus, even a few years later, when hairstyle among black women and men started going to natural, I kept pressing my hair. I didn't want my husband to say that I was too proud to press.

Nothing, however, upset me quite as much as the time he filled me in about his exploits with his ex-girlfriends. We used to talk intimately about many subjects, which I had done with my parents and which I assumed was as proper with a spouse. But the intimacy I had grown up loving consisted of sharing with loved ones our own personal thoughts and feelings, joys and fears, not the bedroom secrets of others. In fact, on the day in question, it took me more than a few moments to realize we were in an area in which I was very uncomfortable. We had been just talking casually, as we often did, when he somehow segued into the subject. Suddenly I was hearing about how he had had sex with three of his ex-girlfriends while he was still at UCLA, naming them all. With one of them, he described how he used to go with her to motels, during afternoons when they were supposed to be in class. With another, they did it in the back seat of his car. He was very explicit about the kinds of noises each made and the kinds of things each said during orgasm. He made a point of telling me how one of them had cried while making love.

Raised in a family that respected every person's right to privacy, I was stunned to be hearing these intimate details of other people's lives. I didn't know what to say, so I said nothing. I tried to rationalize that perhaps he felt that he was just being open and honest with me about his past, but why the graphic details? This was a new side of John I saw that day. He was as capable of being tacky and crude as charming. More important to our future relationship, it raised a wall between us. "He doesn't have any respect for these women," I thought, and almost unconsciously, I started to keep private feelings from him, always aware that I might inadvertently reveal something about myself that might come to be bandied about with others. After what I had just heard, how could I trust him not to share with someone else my own intimate secrets?

Not long after this tale-telling session, John and I went to a party. A couple of the ex-girlfriends—who were still part of our social circle—were there. Each time I looked at one of them, all I could think about was the way that she reached orgasm. There I was standing at the punch bowl and identifying each of them by the different noises they made. It wasn't that I was laughing at them. In fact, I recall being embarrassed about where my thoughts took me, and struggling to force out of my mind these intimate facts about them that I should never have known. The more I struggled, the angrier I became at John. He had not only violated the privacy of these basically decent people, he had put a burden on me in any possible relationships I might develop with them.

John and I had been counseled by our minister not to have children until after he graduated from law school and was well started in his career. We listened to the minister's advice and agreed that he was right we would wait. But it didn't happen that way. In the spring of the first year of our marriage, which was my third year teaching, I became pregnant. I had the best fourth-grade class ever that year; the children were bright and eager and in every way absolutely wonderful. I had asked the principal if I could teach the same group in fifth grade. She told me I had to write a letter to each parent, and if they all gave their consent, it would be all right. Every parent signed on. It was only after it was all arranged, that I discovered I was pregnant. The school system at that time required pregnant women to take a leave without pay beginning their sixth month, so after all the trouble I had put everyone through, when I came back to school the next September I could only teach the class through October.

When I first discovered I was pregnant, despite the timing, I was happy. I flew home from the doctor's office and told John that we were going to have a baby. I'll never forget the look of both surprise and happiness on his face. I told him the baby was due in January. We sat down and figured out how we could make it on our savings and his part-time job at his father's insurance company. It would be a struggle for a while, we knew, but we were both so excited about the baby that we were willing to make the sacrifices.

We agreed that if it was a boy, we would name him Johnnie L. Cochran III, after John. If it was a girl, I could pick the name. He hoped it would be a boy, I think, but I hoped it would be a girl. I didn't know if I would be able to raise a boy very well—I'm so much the frilly, feminine type.

Josie was the first friend I told the good news. She took me over to visit her aunt, who had just had a little girl. This baby, named Melodie, was the most beautiful child I had ever seen—tiny, graceful, cute, and smart, and I decided then that I wanted to name my baby Melodie.

I wasn't prepared for the downside of pregnancy—morning sickness every day for nine months. When I was teaching, it was okay, maybe because I was too busy to focus on how nauseated I felt. But after I was forced to stop teaching, I felt terrible every morning. And sometimes it wasn't just in the mornings. John and I were going to a party one night and I had borrowed a yellow dress from Yvonne for the occasion. I promised to get the dress back to her right after the party. But it took longer than I expected. At the party, I suddenly got sick and Yvonne's dress suffered the consequences. I took it to the cleaners, of course, but Yvonne still hasn't forgotten the incident.

I was bored to death those last few months of my pregnancy. There was nothing to do except stay home and try to take care of myself. I ate a box of crackers a day, to help with the nausea, as well as anything else I could find in the house, and ballooned upward fifty pounds.

One event that eased the boredom was the baby shower that was given for me by Juanita, a fellow teacher who lived nearby and with whom I had often ridden to work. There I was again, the object of attention at a large gathering of women bearing gifts.

Men may find these women's rituals silly, but they are a time and a place where we pass on to each other knowledge about life's ups and downs and how to get through them. I heard stories about labor and giving birth and received many tips on taking care of babies. Not only did I get all that advice, however, I also got just about everything we needed for the baby. Josie gave us a crib and bassinet. Someone else gave us a diaper service. Another person gave us a supply of baby food. Our friends knew things were tough for us financially, and they were very, very good to us.

I was due in early January and, at last, New Year's Eve rolled around. There was a party with some of John's law school friends that we were supposed to go to. We showed up quite late and everyone there was so surprised when we walked in. When we hadn't shown up on time, or close to it, everyone was certain that we were at the hospital.

I was kind of hoping that the baby would wait until my birthday, January 14. I had gone on this long and I figured I could go on a couple more weeks. On the night of January 5, I played in my first bridge tournament. My friend Myrna and I were partners. Novices, we played terribly, but were having fun and stayed until late. I got home about 1:30 in the morning and John was already asleep. Just as I came in, I felt my first labor pains. I woke up John and frantically started throwing things into a suitcase. The labor pains were coming closer and closer. I was worried we wouldn't make it to the hospital, which was all the way across town. John got us there, however, in plenty of time. It turned out that I was going to have a very long labor.

At 2:30 the next afternoon, I was still in the labor room, agonizing away. I don't ever remember being so tired and sleepy. John was sitting there with me. Suddenly, I saw this tiny foot coming out of me. I screamed to John, "Get the doctor. It's her foot!" I pushed the foot back as best as I could and tried to keep calm. John ran for the doctor. When Dr. Weiss arrived, he said, "My God, it's a breach." I was terrified. All I could think about was how my parents' first baby had died—what if that happened to us? Dr. Weiss started to work immediately. John was told to leave the room. As they gave me anesthesia, I drifted off imagining John pacing frantically around the waiting room. I'm not sure what the doctor did, but within a few minutes, a beautiful, perfect little girl was born. We named her Melodie.

It was a very happy time when we first brought that beautiful little baby home from the hospital—even if she did wake us every few hours every night. She was a good baby—once we got her feeding schedule under control. I took her everywhere and introduced her to shopping at an early age, as well as to the movies.

About six weeks after Melodie was born, I had an appointment with the gynecologist for a checkup. John agreed to watch the baby while I was gone. The round trip to Beverly Hills to the doctor, the office wait, the exam, and a short stop at the drugstore took about three or four hours. When I got home Melodie still had on the same diaper that she had on when I left. It was soaked several times over. The dried milk trailing down the side of her mouth was exactly as it had been when I ran out the door. John had never changed her or even picked her up. He may have forgotten that she was there with him. He had the old-fashioned male idea that taking care of a baby was women's work, that women enjoyed changing diapers and didn't have the same aversion to it men had. I decided that Melodie would be better off if I hired a baby-sitter in the future.

John was much better at studying law than forcing himself to change diapers. He graduated from law school the June after Melodie was born. At the celebration after the graduation ceremony, all I remember is John and a bunch of other exhausted law school friends standing around worrying about the bar exam, which they would be taking in a few months.

John studied hard for the bar exam. He took bar review courses and went to study groups with other students. He was a complete bundle of nerves as the day of the exam approached. The California bar exam is one of the toughest in the country. Not only could he not be sure he would be allowed to earn his living as a lawyer until he did pass, taking the exam again was not an ordeal he wanted to repeat.

After the bar exam was over, John still had to wait several months to get back the results. When that fateful day neared, his nervous energy shot off the charts again. The morning the results were released, he got up at an ungodly hour and rushed off to see if his name was on the list of those who passed. I got a call just after the results were posted. Filled with wild excitement and relief, he told me that he had passed.

I was just as relieved as he. Anyone who has lived with someone studying for the bar knows why I wouldn't want to relive the experience. I told him how proud I was of him and what a great lawyer I knew he would be. He asked me to call his mother and tell her that he had passed. When we hung up, I called John's mother and told her the good news. Surprised to hear from me, she said, "I know. He just called here." He had been so wound up that he had called her before calling me and then had forgotten that he had called' her.

John's first job after passing the bar was in the Los Angeles city attorney's office. He wanted to go into private practice eventually, but now was not the time. He needed some experience under his belt, and some fraternal contacts, before he took the financial risk that hanging out his own shingle would entail. I also went back to work and we began to save money so we could buy our first house.

We were still so much the model couple to all our friends, everything going so well in our lives. Even Josie—who was Melodie's godmother and doted on her—must have thought that perhaps she had been too hard on John before the wedding. I didn't ask her, but I figured that now she probably agreed with me that I had done the right thing after all. She had to agree that while

61

there may have been some things about John that weren't perfect, he was a good man and a responsible and loyal husband. By now Josie knew firsthand that finding a good man you could be happy with was not as easy as it seems to daydreaming young women. Josie had had some whirlwind affairs, but no man seemed willing to accept the independent-minded Josie the way she was—and she wasn't going to change for anyone.

When Melodie was about a year and a half, a tragedy occurred that shocked and saddened me. I got a call from Auntie. Her daughter Doris had been shot. Doris, by this time, had divorced her first husband, Hillary. She had moved to El Paso, to be near Auntie, and had met and married a black police officer. They had a young baby, but Doris decided to leave the policeman—his work was stressful and he was the kind of guy who came home and took out his stress on his wife and kids. Doris took her children and moved in with Auntie and Uncle Raymond. Twice hurt by men she had relied upon, she decided to go back to school and learn a skill she could use to support herself. But the policeman was not one to let go easily. Again and again, he begged Doris to reconcile with him, and when she refused to do so, he alternated between promises that he would change and threats that she would be sorry if she continued to reject him. Today we recognize the pattern as that of a batterer-stalker, but in those days people had only their own experiences to educate them, and Doris could not imagine that her husband would do real harm to her, his wife and the mother of his child. One day he came by and asked her to go for a ride with him, just to talk. She consented, and brought their baby along. He again pleaded with her to get back together with him. But her answer was firm—no. He brought her back to Auntie's house and, as she was opening the door to go into the house, he pulled out his gun and shot her, then drove away.

The bullet went into her neck. Auntie rushed to the door at the sound of gunfire and found Doris lying in a pool of blood, still clutching her baby. Doris was alive when she got to the hospital, but went into a coma.

I left Melodie with John's mother and flew down there. As I sat at my beautiful cousin's hospital bed, I prayed for her. It was the first time I was aware of the horrible violence toward their wives that some men are capable of. It frightened me. I thanked God that the man I married wasn't the violent type. I stayed a week, trying to comfort Auntie and help her with Doris's children. Then I had to return home to Melodie, John, and my work.

Doris came out of the coma after three months but remained partially paralyzed from the injury. She had to learn how to talk and walk again, and Auntie arranged for her to have therapy. With her mother's indomitable support and encouragement, Doris made a good recovery, though she was never quite the same. She speaks more slowly and has restricted movement in one leg and one arm because of the injury.

What happened to Doris had a very deep impact on me. After that, if a girlfriend ever told me that her boyfriend or husband had behaved violently toward her, I would tell her to leave him. "Don't think about it twice, because you may not get a second chance," I'd usually say. I didn't suspect then that before long I would be faced with the same choice myself.

John and I had saved a sizable down payment and we began looking for our first home. We found a cute Spanish deco house on a street called Sutro in a mixed area with blacks, whites, and Asians. The schools in the neighborhood were good and the street had other pretty homes with nice, manicured lawns. The house itself had a large sunken living room, three bedrooms, two baths, and a two-car garage.

I thought the house was just right for us. We made an offer and it was accepted. John and I had fun decorating the house and buying new furniture. Life seemed to be going as planned. We both shared a drive to get ahead, a desire for upward mobility—

although the amount of upward mobility each of us wanted and the price we were each willing to pay for it turned out to be very different—and buying this house was a big step on the path to success. I thought wistfully of how proud my parents would have been if they could have seen how well we were doing.

For John, though, the house wasn't quite enough to demonstrate our budding prosperity to the world. Wanting to look like a successful man as he drove back and forth to work, he set his heart on owning a new Cadillac. A fairly new lawyer, he hadn't established credit, so he couldn't get a loan. He asked me to go to my teachers' credit union and secure the loan for his car. I did, providing him with his first entry of many to come into the luxury car world—a drab, olive green Cadillac.

While I obliged him by getting the loan, I never liked Cadillacs. I thought of them as pimp mobiles, an attitude I probably got from my mother. But it was bolstered by my own experience whenever I drove the car. I recall one particular time when I had to borrow the Cadillac because my car was in the shop As I drove to work, men whistled at me and followed me down the street. I was so happy to get my little yellow Mustang out of the shop and told John that I never wanted to drive that Cadillac again. But he loved it. He loved being noticed.

John never planned to stay very long at the city attorney's office. His constant thought was about the day he'd be able to go out on his own. We often talked about how black people, in order to get a fair chance in the justice system, needed good black lawyers—lawyers who understood firsthand the discrimination they faced. There were a few black attorneys in L.A. who had done really well, and these became John's role models. One was a man named Charlie Lloyd, an attorney with a powerful reputation as a criminal lawyer. In the black community there was a saying then, "If you've done something wrong, don't call on the Lord, call on Lloyd." Another role model and sometime mentor to John was an older, successful black lawyer named Bob Robeson who had built up a booming private practice, becoming quite wealthy in the process.

With each passing day, John became more and more impatient to leave the city attorney's office, picturing himself taking his own place among these respected and successful black attorneys in the community. His opportunity came when a black lawyer with an established practice, Gerald Lenoir, offered John the chance to come in with him. John was guaranteed a base salary and a percentage to take Lenoir's overflow work—mostly personal injury cases and a few criminal cases. Gradually John began to attract clients of his own. Finally, arranging to share space with two other young lawyers, he decided to make the leap and open his own office. I was thrilled. I knew he felt as my father had felt when he obtained his first deejay job, that he was at last living his dream.

But while John's career dream was coming true for him right on schedule, our dream marriage had only a short life left to it. It soon became a nightmare.

Chapter 6

On August 12, 1965, John and I looked out to find frightening columns of black smoke, which we could smell as well as see, curling high up into the sky over Los Angeles. According to television reports, a massive riot had broken out in Watts, a few miles to the south of us. Worse, it did not seem to be a passing outburst of a few kids with nothing but time on their hands. A rather broad cross-section of the community seemed to have joined in, and there was such a frenzy in the activities of the rioters many people feared that things would get totally out of hand and consume the entire city. Each day, the fires seemed to be coming closer and closer to where we lived. We didn't go out for several days, sitting either in front of the TV trying to extract some news or on the telephone, worried not only for our own safety but for the safety of people we knew and for our city.

John paced around the house, angry about the situation. "It's a shame that blacks have to fight for their God-given rights by destroying things—particularly their own property. If we could get justice through the courts like everyone else, this wouldn't be happening." I agreed. Centuries of injustice, not just in the courts but in many areas of life, had finally ignited these flames all around us. But I mourned the turn toward destruction and violence as a means to fight injustice. By the time it was over, thirty-four people were dead, twenty-eight of them black, and the property damage—almost entirely in the black community—was over $40 million.

Watts provided the earliest clue accorded most Americans—whites and even middle-class blacks—of the extent of the rage simmering within our inner cities. But it was only the first of many ghettos across the country to go up in flames that year and in the next few years. This was the sixties, however, and it should be remembered that it wasn't just black rage that was suddenly expressing itself in violence. Conflict seemed to be erupting everywhere. Masses of irate college students marched on campus after campus to show their opposition to the Vietnam War and the draft. At the Democratic National Convention in Chicago in 1968, the police turned on student protesters so aggressively that a commission report later characterized their behavior as a police riot.

Within a short time, college campuses across the country were settings for confrontations between university administrators and students. There were building takeovers and even a few bombings. At Kent State, National Guard troops answered a massive student protest with gunfire, resulting in the deaths of four people. Some commentators traced the country's sudden turn toward violence to the brutal assassination, in a public square and in front of his young wife, of our president, John F. Kennedy, a leader whose words, if not always his actions, had inspired new hope in many Americans.

When anger started to tear apart my marriage, a little more than a year after Watts, I wondered if the rage John was directing at me might not be part of this same chain reaction of rage that was burning its way across the country—frustrated folks lashing out senselessly, even if it meant taking out their wrath on their own people.

Yet it must be said that unlike many of the despair-filled people rioting in the nation's ghettos, John didn't have that much to be angry about. His new practice was off to a good start. He was getting cases and earning money. From the very beginning he tried to build a general practice—civil, criminal, personal injury, and even some entertainment law. I have often been asked how a lawyer with a predominantly black clientele could have found such financial success so quickly. The first part of my answer is that John always had white clients as well as black. But it is also important to remember that all blacks are not poor, that even in the sixties there were blacks earning good money.

It is certainly true that at the time John was starting out, there were many, many African Americans who had not yet been allowed to participate in the gains of the civil rights movement. Yet it is also true that many others had managed to make themselves part of the rising tide of American prosperity, and these new business owners and professionals needed competent legal representation to help them hold onto what they were building for themselves and their families.

Still, John's specialty was criminal law, and he was fast developing a reputation as a skilled trial attorney. He'd often ask me to come and watch him in the courtroom, and when I had a day off I sometimes did. He had a cross-examination style all his own. Many trial lawyers ruthlessly attack the other side's witnesses, trying to catch them in narrow contradictions or break them down so that they become belligerent or confused. But this wasn't John's way. He was afraid that if he leaned too hard, if he stripped a witness of his dignity and humiliated him, the jury's sympathy would shift to the beleaguered witness. And this sympathy could easily become a grudge against the lawyer, and ultimately against the defendant. John's special style was to be friendly, respectful, and patient, carefully cultivating the idea that even a witness who might not be an outright liar could be in error in his recollection of certain key events.

But he was most brilliant at summary. Patiently, he'd recount all the evidence that had been submitted to the jury in such a way that it added up to the inescapable conclusion that there was more than one way a reasonable person could see things, opening the door to the criminal defendant's best friend, reasonable doubt.

These summations were delivered in so relaxed, so folksy a style that they seemed totally extemporaneous, as if he were just thinking it all up as he went along. But on occasions when I was in court, they rang entirely differently to me. I had heard them rehearsed the night before, as John tried them out six ways to Sunday, until he was confident he had it just right. Often, his final decision relied in part on whether this approach or another seemed to get from me the reaction he would want from the jury the next day.

If I had been in court to hear him deliver a summation, he'd corner me that evening and demand that I confirm for him how great he had been. "I mesmerized them," he'd say. "Don't you think?" And I'd have to answer that yes, he had mesmerized them.

There was another aspect of his work that troubled me. It seemed to me that many times his clients had obviously committed the acts they denied committing that I wondered why John himself wasn't troubled. I'm not a lawyer, and it goes without saying that lawyers would think my reaction unsophisticated, that every person accused of a crime is entitled to legal representation that takes into account his interests only. But within my lay sense of things, the obligation to provide legal representation did not necessarily include an obligation to cloud the facts of the case. Whether the client did the deed was really a different question from whether or not he was guilty of a crime, and if he had, what his punishment should be. Wasn't there a proper role for lawyers that did not involve manipulating the facts—making sure that judge and jury saw the acts of the client in proper context and took into account all extenuating circumstances, including whether or not there had been criminal intent. "The devil made me do it," can be an honest defense. "The devil did it," cannot.

But John, like most successful defense lawyers, thought about winning, about getting his client off. The more outrageously guilty the client, the more an acquittal elevated the reputation of the defense lawyer who had won it. To pull off these tough wins, John was learning all the tricks in the game, learning to throw up smoke screens and lots of legal mumbo-jumbo which, in my opinion, didn't enlighten the jury but confused it—stuff that had nothing to do with guilt or innocence but only with the need to create a reasonable doubt in at least one jury member's mind about what really did happen. This was how, like most criminal defense lawyers, he saw his job, what he saw as his responsibility.

As I left the courtroom on one occasion, I hoped John would one day soon have an opportunity to put his superior talents to work representing an innocent client. Such an opportunity did come to him. The Deadwyler case made John a star. And in doing so, changed him forever.

Just months after the Watts riot, Leonard and Barbara Deadwyler were riding in a friend's car, when Barbara, pregnant at the time, felt a stab of what she thought was a labor pain. Leonard tied a white handkerchief to the car's antenna before he took the wheel and sped off for the hospital. Within minutes, several police cars were on his tail. He pulled over. One of the officers got out of the patrol car and approached. Apparently, the officer didn't like Leonard Deadwyler's explanation for his vehicular haste, because he poked a gun through the car window and shot the man dead.

The pistol report reverberated through the black community, still simmering after Watts. Despite police contentions that Deadwyler himself had provoked the gunfire, we all knew better and were infuriated, first because the police had shot and killed an innocent and unarmed man and then because they had shown themselves so willing to sully a dead man's good name rather than take responsibility for their improper behavior. The next morning John was called by the family. Would he represent them?

The case went to a coroner's jury. Because of the tension throughout the city at the time, the inquest was televised live, and for the first time John had the big stage he needed, and a wide

70

audience for his charismatic style. Before this case, John had been like a theme looking for the right movie. Now he had found the script. And he played his role to perfection.

Even now I want to believe that it was not only the slew of positive publicity the case brought him that so thrilled John, but as well that it was a chance to make a difference for black people. He was not representing just another criminal, but the good name of a man who had behaved properly and responsibly in a crisis, one whose life had been taken from him simply because he was in such a hurry to get his pregnant wife to the hospital that he didn't take the time to show proper obeisance to the white police officer who pulled him over. Most disturbing was the hard reality that Leonard Deadwyler would not be dead if his skin were another color. Despite the clear affront to justice, there were many who had doubted there would be a fair hearing because the offending officer was white and Deadwyler had been black.

Because of the special rules of a coroner's inquest, John was not allowed to ask questions directly but had to submit them through the deputy district attorney handling the case. However, he had such a sense for the theater of it all that he turned these constraints to his own good, giving the television audience a picture of him whispering in a public official's ear, and then having that public official ask his questions: "Mr. Cochran would like to know …," or "It bothers Mr. Cochran that" such and such was said, or done.

The case was exactly what John needed to establish himself as the black Clarence Darrow. His superb performance throughout the case exceeded even his own expectations, and soon all of Los Angeles was describing its new legal star in glowing terms.

In the end, sadly, no charges were brought against the police, but at least they had not been able to have their own version of events, unchallenged, become the public record, as had always been the rule. The case also gave John his first taste of fame. The picture of a black lawyer fearlessly challenging the system, despite being prevented from speaking openly to the jury, made John a new hero for L.A.'s black community. There was no telling how high such a man could go.

As for me, I still loved teaching. It was my own way of trying to make a difference. About this time, Black History Week was instituted at my school. My first reaction was to ask if this meant that the rest of the nine months we would continue to restrict ourselves to teaching only white history. But I soon got into the swing of things. The problem was the curriculum. The administration wanted me to teach the students about Charles Drew and Phyllis Wheatly. I wanted to tell them about Smokey Robinson.

I spent a day on Wheatly and Drew and then asked the students to draw their chairs close together. And I began the story of a very thin young boy named Smokey Robinson. Smokey, I told them, lived in the Brewster projects where those who would grow up to become the Temptations and Diana Ross lived. The other boys would not let Smokey play football with them because he was so small and thin. For hours at a time he used to stand by the window and watch the other boys play.

One day he started to write poems about his feelings and then set them to music. Moved by Smokey's story, the children soon started talking about their own experiences. On the last day of Black History Week I brought in all those great Smokey Robinson albums and, of course, as soon as Smokey was belting out his songs and the walls were vibrating, the principal came into our room. In my box later that day there was a note: "See me before you go home."

I explained in my defense that these children needed to hear about someone to whom they could relate, someone who felt as hurt about being left out as they felt. This was a way I could introduce them to the voices that come out of oppression. The principal smiled. She understood. And I ended up with a promotion instead of a reprimand.

I still thought then that life was sweet and optimistic. I was the proud wife of a brilliant, young black lawyer, a rising star in the community, the mother of our beautiful daughter, and a successful teacher. Despite the anger and violence in the air, we were still living the dream that Martin Luther King had bequeathed us.

Soon after the Deadwyler case, John began to work harder and harder, or at least longer and longer hours. Instead of coming home for dinner, he would call and say he would have to work at the office until 8 or 9 p.m. I was disappointed, but I accepted this as the price for being married to an ambitious man.

Eight or nine in the evening soon became 10 and 10:30 and then 11 and 11:30. I didn't think most lawyers practiced law at 11:30 at night, not night after night. But then again, I understood so little about being a lawyer, so who was I to decide when those mental machinations required of lawyers could best be turned on. Disturbing possibilities occasionally came to my own mind, but I felt I should not be suspicious. All I ever wanted was to be a good wife, and I knew that mistrust was the end of intimacy. But one night, when we were sitting in the den, I decided to say something. As casually as I could, I asked him, "John, are you telling me the truth about all of these meetings at night when you don't get home until 11 or 11:30?"

"Oaww," he said, Of course I am. "I'm trying to build a successful practice, and it just takes longer to do things right than to just leave the office at a certain time. Don't I always come home?"

I really had no reason not to believe him. John's practice was certainly building. And when we were together, he seemed as devoted as ever. I tried to put these suspicions out of my head. Still, the nights were long and lonely. John rarely was home for dinner or even in time to put Melodie to bed.

Then he took his first weekend trip, to see, he said, a client in jail in San Francisco. He couldn't afford to take the time out of his daily schedule to fly up there, so it had to be done on a weekend. It would be "too boring" he said for Melodie and me to come along and sit around while he visited the prison.

Stupid me. I might have gone on believing him forever except that this time he was unlucky enough to get caught. The next Tuesday a friend called and asked how I had liked Vegas. "Vegas?" I asked. Yes, she had seen Johnnie up in Las Vegas Saturday night. I hung up, upset, very upset. Damn him, I thought to myself. How could he do this to me? People on their way up to a prison in San Francisco don't get lost and find themselves in Las Vegas.

I was angry and hurt and even a bit ashamed. Did he value what he had with me and with Melodie so lightly that he would so cavalierly risk it all for a weekend romp? More important, if he had deceived me in going to Las Vegas, had he been deceiving me all these months when supposedly he was hard at work at his desk until nearly midnight?

After a couple of hours I realized that I would have to make an important decision—what to do next. Did I confront John with what I had just learned? Was *I* ready for that? What would I do or say if the whole ugly truth of these past months came pouring out? The alternative was to say nothing. Did I lie to myself and make believe I had not received that call? Did I go on as things were, pretending that I was married to a loving faithful husband? I played out each option in my head, again and again, until the decision was made: I decided to take the risk. I confronted him that night.

If I expected contrition, maybe a request for another chance, I guessed wrong. Without missing a beat, he went right into legal mode. Who me? Wonderful Johnnie? Maybe your friend saw someone she thought was me—there are other handsome black men in America, you know—but I was in San Francisco. When I didn't look persuaded, he turned to a favorite defense tactic, attacking the credibility of the witness. "Your girlfriends are silly people, hardly trained observers. I don't see why you would believe anything they have to say. You should believe your husband."

I'm sure some women would have told themselves, "Yes, yes, he is my husband, I owe it to him to believe him." But I didn't believe a word he was saying.

The long nights "at work" continued. If I brought up the subject of his late hours, even in our most relaxed moments, he immediately took the offensive, saying I was crazy, that he wasn't doing anything or seeing anyone and that I was letting busybodies wreck our wonderful marriage. But he still didn't come home at night.

It got worse when the basketball season started. John would say he was going to a game with the guys and wouldn't get home until 1:30 a.m. Now everyone who owns a TV knows that basketball games don't end at 1 a.m. But I didn't even ask where he had been. I got to the point where I didn't want to be lied to. I'd been through it so many times already that I knew the pattern he'd adopt if I did ask, that the best defense is always a vigorous offense. I was crazy; my friends were jealous of my wonderful marriage. He was the best thing that had ever happened to me. As many answers to that one simple question—where have you been?—as he thought necessary to confuse the issue.

But if I was crazy, I wasn't stupid, and my brain was telling me that my growing suspicions were on track. On the emotional front, however, I was a wreck. I couldn't focus on the things I really should have been focusing on—my work, my little daughter. My obsession about being jollied by L.A.'s most persuasive lawyer began to consume my whole day. I awoke thinking about it. I went to bed thinking about it. Was I blowing some minor thing way out of proportion? He had accused me a dozen times of being a little schoolteacher who didn't know the ways of the world and needed to imagine things that weren't really happening to spice up my life. Or was I just facing up to the fact that the man I had married was so skilled at getting people to believe him that he could sell an ice-cube maker to a North Pole native. The more skillfully he deflected my questions, the more I kept thinking to myself: He brags about mesmerizing juries. Why should he be any different with me?

I needed help, but who could I turn to? I was too humiliated to tell even Josie. There was only one person I could turn to—Auntie. Tearfully I called her in El Paso and told her that I suspected John of being unfaithful to me. I was surprised by her advice, although looking back on it now, I realize she gave me progressive advice for that time.

"If you really suspect him, but you don't want to break up your marriage over nothing, then hire a private detective and find out for sure."

Hiring a detective turned out to be surprisingly easy. You find one in the yellow pages. The detective asked me to describe John, to describe his car, to tell him where we lived and where John's office was. He asked me where I thought John would be at certain times of the day so that he could pick up the trail.

Within days I had my answer. John was indeed going to basketball games. But after the game he went to the home of a Patricia Sikora, who lived in Eagle Rock, a white suburban area. The detective said he got there at 11:30 and the lights went out at 11:45. The lights came back on again at 12:45 and he left her house by 1 a.m. There were a couple of games each week and he followed the same pattern each game night. That was John, all right, always a creature of habit. It also fit in with what I knew, that I could expect him home by 1:30.

In my generation, marriage was the thing you lived for, to marry a good man, a loving man, and to have his children and live happily ever after with him. How could he do this to us—to Melodie and me? How could he do this to himself?

That day I learned the value of keeping my own secrets. I didn't say anything to him that whole week. I put the report from the detective in the glove compartment of my car. I drove around, sometimes in a rage, other times crying. I yearned for my parents. We had talked about everything. What would they have said to me now? But, of course, there is no training to prepare you for your first betrayal. Each of us has to deal with it the best we can, with whatever resources we're lucky enough to have.

To the outside world—to our friends and acquaintances—John and I seemed to be the same model couple, talking and laughing about things. But a debate was raging within me over what to do about my marriage. Of course, John was not home enough to notice that I was not my old self.

Every day I didn't confront him with what I knew, I felt myself being caught up tighter and tighter in his web of lies. Yet I cringed at the prospect of confronting him, knowing that if I tried to, he would go on the offensive and become abusive, and I would be no match for him. I had always hated confrontations, and I expected that he would have no qualms about using knowledge of these feelings of mine to back me down. I tried taking it one day at a time, but every morning I'd wake up and remember that this life I was living was a fraud, not at all the life I had planned for myself.

When John tried to be intimate, occasionally I went along, but with no feeling, no heart, and no real pleasure. Once I tried to bring up the subject indirectly, telling him I felt he didn't respect me and the institution of marriage. Good old John. "Oaww, I respect you. I respect marriage. If I didn't respect you, I wouldn't be here."

I obsessed some more, and finally I steeled myself and spit it out. His response was predictable. The defense attorney went into action putting the accuser on trial. "How dare you go to a detective!"

When I didn't cower, he shifted gears as smoothly as only he could, telling me the detective was lying. They were all ex-cops. First they decide if they think the person is guilty, then they find the evidence to support their conclusion. When even that tack didn't work he took another side step. The detective was feeding me whatever information he believed might induce me to give him more money, and fool that I was, I was actually paying someone to lie to me.

A couple of years earlier he might have reached me. But not

now. This was his oldest friend he was feeding his snow job. "No John," I said to him calmly, "you're the one who's lying to me. It's not the detective."

A vein bulged out of his forehead. "You're crazy," he said.

It was incredible to me that he would still lie, in the face of the detective's moment-by-moment, blow-by-blow description of everything that was going on other than the actual sex act. But he followed the same advice he had so often given his clients. Admit nothing. Deny everything.

After that, we began to argue over everything. When I asked for money for household items, he would accuse me of spending too much. I told him I thought he was feeling tension about other things and taking it out on me. He would say, "Oaaw, you're crazy. I'm not taking anything out on you. I don't even know what you're talking about."

His disposition was rapidly changing. It took nothing, absolutely nothing, to start an argument. Was he in some tiny hidden part of him guilty over the destruction of our marriage? Or was he frustrated because he would have liked to dump me but was afraid his professional reputation as defender of African Americans might suffer if he left his wife for a white woman as soon as he amassed a little fame and money?

I desperately wanted to put our marriage back together again, but I didn't know how to start that process. I had always been known as a TCOB type, a take-care-of-business person, but here I was in the middle of the most important crisis in my life and I didn't have a clue as to where to go from here. I wasn't myself anymore. I was empty, depressed, unenthusiastic about life. Each day passed, kind of like the weather in Oakland, where every day brings the same thing—fog, gloom, and dank murkiness.

John, of course, had a different perspective. To him it was always, "Why are you complaining?" He would point out that I had a beautiful home, beautiful clothes and, of course, I had him.

I was an ingrate. I should not tempt the fates when they had been so kind to me.

And yet the emptiness inside me told me that whatever it was I had, it wasn't right. Perhaps it would have been enough if this new stage in the marriage had been fine-tuned to include my needs. But everything was always about his needs. We—he and I—had to serve him—Johnnie Cochran the lawyer—so that he—Johnnie Cochran the husband—could provide for us. But I felt that I deserved much more. I deserved a relationship that was meaningful, where both of us were growing in the same direction and trying to accomplish things together.

Instead, we were at a fork in the road. I was going in one direction, looking for warmth and intimacy, but he was clearly headed up, and on his success ladder I was only a low rung to climb up from. I had to be there for whatever banquets he needed to attend with a spouse, to be at his side, smiling, appearing so happy about everything. I provided him with a reputable past and present that he could use to purchase a better future.

I started living in my own little world, trying to insulate myself from him, trying to protect myself from the man who slept beside me in my bed. The only things that continued to give me happiness were Melodie and my job. One night after I had put Melodie to bed but before we turned in, we began arguing. I don't even remember what the argument was over. Our voices got higher and higher, louder and louder. Both of us were at the breaking point. It happened in a flash.

He grabbed me and pushed me against the wall. Anger was oozing out of every pore of his body. Before I realized what was happening, he took hold of my dress. He was tearing if off my body. I was screaming, "Stop it, stop it! Let go of me." He was out of control. Pulling the garment, tearing it to shreds. I didn't know what he'd do. Rape me? Kill me?

Maybe it was the sight of the ripped and tattered dress that brought him to his senses. He dropped the torn garment as if it were hot and turned and stormed out of the house. I stood there, stunned. Would he try to hurt me when he returned? Should I stay? Should I go?

This wasn't the first time he had hit me. It was the second. One night a few years earlier, when John was still working in the city attorney's office, he announced that he was going to a banquet and didn't want me to go with him. He said that only lawyers and judges were going.

"I've never heard of a banquet where there are only lawyers and judges," I said. "Usually they bring their spouses."

I was sitting in the bathroom arguing with him when he suddenly just hauled off and slapped me hard on my face. "You're not going," he shouted. "I'm going. And I'm going alone."

Wild surges of emotions overwhelmed me as I heard that door slam—fear, shame, and anger. He had no right to hit me. I was also puzzled. It was hard to believe that this nice, sweet man of words I had married was capable of violence. I finally went to bed, and when he came home, I pretended to be asleep.

The next morning he was all contrition. As it turned out, he said, there was really no reason I couldn't have gone to the banquet. For him the issue was over, as lawyers say, moot. But a nagging question remained. Would this become the pattern? Would I end up the way my cousin Doris had, badly injured by a man who professed his love for me? How long do you wait before you discover you have waited too long?

Up to now, John had never repeated the violence of that night, and gradually I came to treat that first occasion as an aberration. But the night John tore off my dress those same emotions welled up in me, and those same questions. But this time the answer was clearer. I knew I was in danger of being beaten every time things didn't go his way. I knew I would be yelled at and screamed at every time I confronted him with something he didn't want to hear, that I would soon find myself trimming my responses to things, indeed trimming my every action, all my behavior, to avoid triggering a violent reaction in him.

Still, it was hard to let go. I wondered, is this what has become of our dream? Like Watts, was my marriage burning up in hatred and violence?

The third time John hit me occurred a few months later, in April 1967, and this was the most frightening. It was bad enough that I had sacrificed so much of my own dignity by learning to walk on eggshells to avoid angering him. This incident didn't even require a confrontation on my part. He seemed determined to teach me that certain behavior was expected of me, and I had better comply.

John was home early for the second night in a row. I thought perhaps things were cooling off with his girlfriend—or they were having a fight. The phone rang and Melodie answered. It was for me. While I was on the phone, John asked Melodie who was on the phone and she told him it was some man. When I hung up, he interrogated me: "Why do you have men calling you?"

My face may have shown the irony I heard in such a question, but I surely did not express what I was thinking. Instead, I calmly responded factually to the question I had been asked. I was interested in bowling on a teachers' league and had told them to call me when they had an opening.

Maybe he was transferring his own guilt onto me, or maybe he had come to suspect that because he had been sharing so little of himself with me lately I had gone out and looked for male companionship elsewhere, but he didn't accept my answer. Angry, he sat there and fumed about it until Melodie went to bed. When she was asleep in her room, he grabbed me, holding me by one shoulder tightly—it hurt the way he was holding me—and started hitting me on the side of my head with his fists, where my hair would cover any marks. He hit me three or four times, yelling, "I'm going to hit you where there won't be any bruises." I cried out for him to stop. "You have no right to hit me!" I protested angrily.

But futilely. He didn't stop. My head was aching from the pounding, and though I struggled against his grip, I couldn't free myself. In his own time he stopped. Only then did I begin to cry, out of humiliation as well as physical injury. Who was I married to? What kind of contempt did this man feel for me that he would hit me simply because a man had called me about a bowling league?

That night there was no discussion, no more arguing. I cried to myself for a while, facing the fact that if he had struck me a first time and a second, and now this third time, he would keep doing it, and the threshold he needed for his own self-justification would be lowered each time I allowed it. If his unfaithfulness and his lies made living with him painful, his violence now made it impossible. I would not risk having my fate mirror that of my cousin, Doris. I would not continue in a relationship in which I had to fear getting my partner angry. How would I ever explain it to my daughter if I did?

I knew I had to tell Josie. She would understand and would help me. After John left for work the next morning, I called her up, "Girl, last night John hit me. I'm packing his clothes. I need you to help me." She said she'd be right over.

When Josie arrived, we carried John's clothes to the car. On the way to John's parents' house, I told Josie about John's unfaithfulness, not even sparing her the detailed reports from the detective I had hired. She knew the agony I was in and was kind enough not to say I told you so. She just comforted me with words of support. She was one of the few friends I could have called on at that moment.

So many others were caught up, as I was, in all this black middle-class garbage. This desire not to have anything seem wrong with their lives, as if their failure to have a picture-perfect marriage might make them seem more ghetto black than middle class, or might reflect badly on other black Americans. Too many women I knew would just sweep this kind of thing under the rug as long

82

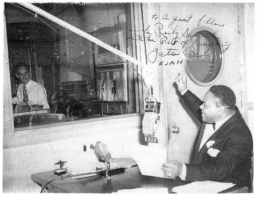

Youree Berry, my father, the first black disc jockey in the San Francisco Bay area, when he was a popular disc jockey at radio station KSAN.

"Auntie" Annette Nelson, my father's sister, who took me in after my parents died and was a major source of strength to me.

Me, soon to be John's second-most prized possession, atop his car.

John and me, during the "magic moments" of our engagement, standing in front of the car I rode in rather than the bus, the fateful day I met John.

John's parents' house in L.A. at the time of our marriage.

John Cochran's parents in 1967—Johnnie Cochran, Sr. (the "Doctor"), and Hattie Cochran. (*Photo: Harry H. Adams*)

As John gives me the wedding ring, my maid of honor, Josie, still looks like she thinks I'm making a big mistake.

We did all the traditional things: cutting the cake, drinking the first glass of champagne, and the bride throwing the garter—but, despite the precautions, the marriage still turned out the way it did!

John and I pose for a photographer at the open house party in 1966—right around the time he started seeing Patty.

Our first home, a comfy Spanish stucco on Sutro Avenue in the Leimert Park area of L.A.

John, Melodie, and me in our first family portrait, during the years after law school when John worked at the city attorney's office.

We celebrated our seventh anniversary in Las Vegas together after our separation earlier that year.

The five-bedroom house on Hobart that we moved to in 1972, where John still lives with his current wife, Dale, and his father.

John was a frequent visitor to this bungalow in the valley—only a few minutes from our house—where Patty Cochran, her daughter, April, and her son by John, Jonathan, lived for many years.

Tiffany and John at her debutante ball. Arnelle Simpson was a debutante at the same ball.

John with all his children, Jonathan, Tiffany, and Melodie.

A family portrait at an award ceremony in 1977. I am smiling because I know this is one of the last affairs I'll have to attend with John.

This time, older and wiser, and with David by my side, I was a content and happy bride.

The family reunites for Tiffany's graduation from Pepperdine University in 1987. I was already married to David and John was married to Dale—and "Life After Johnnie" was a much happier time for me.

as the man paid the bills. They would have never supported me in my decision to leave John, this budding young god to L.A.'s blacks. He wasn't close to the worst kind of wife batterer, these friends might have said; you can't believe that he would ever actually kill you or even seriously hurt you. But Josie understood my needs—I hadn't been raised to be some man's frustration outlet.

When we arrived at John's parents, I told John's mother that John and I were having problems, and I was going to change the locks. I just needed to be away from him. Sadly, she took his clothes and said she thought we all better get together and talk about it. I went back home and called the locksmith. Then I called John's father to ask him to call John. When Melodie came home from school, I told her John wouldn't be coming home for a while, but that she'd have plenty of chances to see him.

That evening a sense of relief came over me. I could feel the tension going out of my body. Little did I realize then that it would be another decade before I would be free of this man.

Chapter 7

The pressure for me to take John back started almost immediately. And it was well orchestrated. So the day after I moved John's clothing out of the house, I went to see a divorce lawyer. A friend of mine, who had been married to a well-known writer, recommended her divorce attorney, a man named Stanley Poster.

I was terribly nervous as I drove to his office. Though my decision to separate from John was initially an enormous relief, the thought of legally acting upon that decision now unnerved me.

What am I doing, I thought? When I recited my wedding vows, they were meant to be forever. I sat in the waiting room feeling ever more anxious that I might be doing the wrong thing. I was also embarrassed—for myself and, yes, even for John—about what I would have to tell this stranger. But each time I thought of getting up to leave, I remembered John's fists hitting my head, and I stayed.

Stanley Poster made me feel at ease. He knew who John was—after the Deadwyler case "Johnnie Cochran," although not quite a household name and not yet a subject for quips on the late shows, was well known locally, especially in L.A.'s legal fraternity. He asked me what happened. I told him about the recent beating and the night John tore off my clothes. I told him that John was having an affair and showed him the detective's report.

He asked me if I was afraid of further violence from John and I said yes. He asked me if I thought counseling would help and

I said no. John, I suggested, would likely ridicule the advice a counselor gave, and I wasn't at all sure he would even agree to go to one. My lawyer didn't ask me much more. He said he would prepare papers for a restraining order and a legal separation.

When I left his office, I felt calm. I knew I was in good hands. I also knew that I had done the right thing for myself. I am not the type of person to wish away problems; now that I had acted, I was looking forward to some time alone to think about what to do next and to regain my peace of mind. I didn't want to jump into the decision about whether to proceed with a divorce or stay in the marriage. A restraining order and a legal separation seemed just about the right step at this stage of things.

As it turned out, I had very little peace. First John called, wanting to come back home. He was willing to talk about our mutual grievances, he said, and he would try to change his behavior in those areas most troubling to me, if only I'd tell him what these were.

"If only I'd tell him what these were!" I was more offended than impressed. This was his wife he was pulling this on, his college sweetheart, the mother of his child, and he was negotiating with me as he would with the other side's attorney. Hearing his voice coming over the phone in this mode brought up intense hostility. "Call me when you grow up. We can talk about it then," I said as I hung up.

Realizing he wasn't going to get very far with me on his own, John enlisted the aid of his parents. One of them soon called me, proposing that we all get together and just talk. It was hard for me to say no to them. In the almost seven years I had been married to John, Hattie and the doctor had become like parents to me. I was very fond of both of them. Reluctantly, I agreed to get together.

A few days later, John, with his mother and father in tow, showed up at the house.

At this first powwow, John's parents tried to assume the posture of mediators, but in their words and actions they were clearly putting forth John's interests. Again it all sounded to me like a well-rehearsed trial. Opening statements were focused on how all marriages have problems and everybody has to try to get over them and go on. For once, John ignored the advice he had often given his clients—that when you're too emotionally involved to plead your own case, you're better off shutting up and letting your lawyer do your talking for you. His very first words upset the mood of respect for me his parents had tried to set. Everything, he said, had just been "blown out of proportion."

I couldn't let that stand. Furious at the implication that domestic violence was something you could make seem less important, I managed to convey to his parents that I had not been raised to be disciplined by my husband, and that I wouldn't tolerate it. "John was hitting me!" I said. "That is the issue."

This message they got. They took their attention off me and turned to John. His father told him he shouldn't hit me. It wasn't right. John listened to these mild reprimands without an argument. But still I heard no apology, saw no sign of remorse or contrition. Not even an admission of guilt. There was certainly no pledge that it would never happen again, for that might be taken as a confession that it had happened in the past.

As this parent-child counseling session wound down, I again took the floor. I wanted to put all the charges on the table. I told his parents that John had also been unfaithful to me. I showed them exhibit "A"—the report from the detective, with its specific details of John's affair. The pained look on John's face as he watched his father read the detective's report made me feel almost sorry for him. But then I remembered that I had not caused him this pain. He had brought it on himself.

I looked into the doctor's eyes, afraid to see the same pain there but expecting that at least the report would earn me some support from this basically decent man. Whatever was in his heart, it did not find its way to his lips. I was stunned to hear him tell me that all detectives lie, and I shouldn't believe what they say. Detectives are working for money. They will just keep lying as long as I keep paying them, he counseled me.

Now where had I heard all this before? Let me think. John's parents' response to the detective's report of John's infidelity was almost word-for-word John's—only without his anger and defensiveness. Then they added a new dimension—I suppose you could call it "playing the race card." They intimated that since the detective I hired was white, he might have been prejudiced against John. I should have known, they explained, that the opportunity to contribute to the public shame of one of L.A.'s leading African-American figures would prove too much temptation for some people to resist.

The more they made these kinds of arguments, the more I began to shut down. I felt betrayed by all of them.

But nobody was really interested in how I felt. "If you guys patch up this thing quickly, people will never know that you've been separated," one of them suggested. That was the final straw. Who were they concerned about here? Certainly not me, I thought. Apparently they didn't care about what his infidelity had done to me or to our marriage. They were more concerned about what "people" would think about their son if they found out his wife had left him. I was disgusted with all of them.

I hated confrontation then and was fearful of expressing my feelings—especially angry feelings—but I excelled at passive resistance to pressure tactics. The more they pushed me that night to get back together with John, the more I dug in my heels and refused to make the concessions I was being pressured to make. "I need more time," I said firmly. They said they'd come by again in a week or so. I agreed to nothing.

When they finally left, I felt both anger and sadness as I closed the door behind them. I never stopped loving John's parents. But I saw that they were blinded by their love for their son and felt that all they had helped him build might be seriously damaged by my hurt feelings. They couldn't possibly be fair to me.

The next day, I called Auntie. Good old Auntie. She said it wasn't right for three of them to come over and pressure me the way they did. I needed someone on my side. She would come up for a visit and the next time they came over, she would be there.

It was a great comfort having Auntie around. I hadn't seen her in quite a while and missed her terribly. Life had been hard for her these last few years, taking care of Doris during her therapy and rehabilitation, as well as taking care of Doris's children. Now here Auntie was again, I thought, taking care of me.

Some days Auntie and I just had a good time together and didn't talk about any of it. We took Melodie to the park or went to the movies. Having been through two divorces herself, she knew there were times when you are sick of talking about your problems. You simply need to get your mind off of them.

Not long after Auntie arrived, John and his parents came over for another talk. One of them began by saying, "This thing has to be resolved right away." Auntie immediately jumped in and objected, telling them to lighten up. "Barbara needs a chance to think things through," she said. "She can't be pressured to reconcile before she feels right about it." They backed off without an argument. Auntie had a strong personality. You either loved her or hated her, but there was seldom any doubt that you were hearing from her exactly what was on her mind, and most people respect that.

After Auntie quashed the issue of how soon this thing had to be resolved, someone wondered out loud if perhaps I was pregnant. I bristled at the suggestion. "Now they just think my hormones are acting up!" I thought. I told them that I wasn't pregnant, trying not to let the hostility I felt show in my response. They droned on, repeating many of the same old songs as the other night—how every marriage has problems, how John was such a good provider for his family, how we both wanted the same things in life and should be together.

Auntie was the only one who seemed to notice that the more John's parents praised their son, the more they pushed for reconciliation, the more I withdrew into myself. She tried to take a middle ground, pointing out that it seemed to her that John and I needed to be talking and she hadn't heard much from either of us all evening.

John took his cue from that and made a little speech about how he realized it was wrong to have hit me. He said he would never strike me again, under any circumstances.

I must admit surprise. John, at least, had come to the point where he was now willing to admit he was wrong. At the last meeting, he would only say that things had been "blown out of proportion." Skillful negotiator that he was he must have decided he'd make his concession in those areas where he knew I would never compromise. Number one, he had hit me. Not once but three times. There was no doubt about that, and I would never agree to have the incidents described in general, euphemistic terms. Number two, it was wrong to have hit me, and that was all there was to it. I would not agree that once the hitting itself had been established we needed to hear from the other side to see if it had been justified. And number three, I would tolerate no more of it. Other things might be negotiable, but these three were not.

But I still couldn't say how I felt about a reconciliation. I could talk to Auntie when we were alone about how angry, humiliated, and betrayed I felt about John's infidelity. But in front of John and his parents I just clammed up. They all continued talking and I tuned them out, thinking about what John had just said, and about what he had not said.

While he had admitted guilt about the violence, what about his affair, his lies? He can't say he will stop the affair because he won't even admit to it in the first place. How poor a foundation for future trust when he continues to deny the reality I know to be true, tells me that a betrayal hasn't happened even though he knows I have proof that it has. As I was warring with all this in my head, John's father spoke softly to me, "Bobbie, you can't keep holding it all in. If you want to talk to someone, you know you can always call me."

Tears came to my eyes. I appreciated his sensitive words but his sensitivity was beside the point. "The one I should be talking to," I thought, "is John, not his father. But we can't talk anymore— not about anything that matters."

At the end of that grueling session, I still insisted that I wanted more time. I knew I had been happy in the weeks apart from John. I wasn't yet sure that divorce wasn't the best answer after all. I knew I would have to make that decision eventually, but I wasn't ready yet. As they were leaving, John said he would call me. "Maybe we can just have dinner or see a movie. We don't have to talk about reconciling if you don't want to." I nodded, "Okay, John," too worn down by the evening to say anything else.

John came by to pick me up about a week later. I had been nervous all week about agreeing to see him alone. He promised me we would just go out and have a nice dinner. And that would be all. At dinner, he talked about his cases. We talked about things going on in the world. If it had really been just a dinner date, I would have seen it as a successful date. We were both cordial, like acquaintances, but I didn't feel close to him.

He told me he had gotten an apartment. He said he didn't know how long I was going to take to make up my mind about reconciling, and he didn't want to continue staying at his parents. I was again surprised. John had lived with his parents up until the time we married. On the way home, he showed me the apartment. It was small and simple. It hardly looked lived in.

That night when I got home, Auntie wondered how things had gone. I told her that the evening was fine, but I still didn't know if John could ever be a good husband. Whenever we talked about my marriage, Auntie always made sure I knew she understood what a disappointment John had been to me and how betrayed I felt. But now she also broke the bad news—she thought that John and I should reconcile. "He's not the worst husband in the world," she offered. She thought I owed it not just to John, but to Melodie and myself to give it another try.

Auntie, like most women of her generation and mine, gave men a lot of slack. If they paid the bills, and particularly if they earned good money and spent a fair share of it on their families, women were supposed to endure whatever they had to for the sake of the children and the marriage. "He will probably always be a liar," Auntie said. "And he'll run around until he gets bored with it. But he wants to take care of you. He wants you to be his wife." What a recommendation, I thought.

When I put Auntie on a plane to go home a few days later, she hugged me and said, "Try to forgive him, Barbara. Try to give it another chance."

Those words reverberated in my head over the next few months. I saw John quite often. We went to the movies or to dinner. We tried to rebuild some sort of relationship. We started to laugh together again. But I couldn't let him touch me. Not yet. I couldn't let myself be vulnerable to him.

I was still weighing everything in my mind. On a couple of occasions now, John had repeated his promise never to hit me again. I reminded him that he had said that before. He responded that this time it was different. He had really learned his lesson. He would never hit me again.

I was beginning to believe he could change—as far as the physical violence was concerned. He respected his parents immensely. Since they told him that hitting his wife was wrong, he probably now agreed it was wrong. Perhaps he wouldn't do it anymore.

Despite the experiences I had seen in my life—particularly the tragic experience of my cousin Doris—I was like most women in that I knew little about violence in marriages, about why it occurs or how likely it is to continue once it starts. This was 1967. No one that I knew talked about "domestic violence" or "spousal abuse" in those days. There were few studies tracing patterns in "wife beating," as it was still called in those rare moments when it was mentioned. Like diarrhea and mental illness it was never mentioned in polite company. I didn't know that there were other forms of abuse that can precede the physical abuse, or occur alongside it, or replace it when the physical abuse stops. All I knew was that John didn't seem nearly as violent as the police officer who had professed his love and then shot my cousin Doris. So on the issue of the violence, I decided about a month or so later to trust that John had read me loud and clear, and I was ready to take my chances. I would have reconciled with him much earlier if that had been my only concern.

John's affair and his lies to me, however, were more troubling. I remembered John bragging to me about how he was sleeping with three women while he was at UCLA. One woman apparently wasn't enough then. I turned out not to have been enough during our marriage, and I now had to face the fact that one woman might never be enough for John. His ego was too big to be satisfied by one woman, I thought.

Before we separated, he had often thrown up in my face how this woman or that woman thought he was handsome or smart or whatever. If there was something he needed that I wasn't giving him in the marriage, I wondered, why had he never talked to me about it? He just found other women who were apparently all too happy to provide it.

I hated his infidelity. It hurt me deeply. He had promised to love and cherish me until death do us part, but this forever had lasted less than seven years, if in fact it had lasted that long. Perhaps teachers have a different attitude toward broken promises than do lawyers. Break your word to a child and you may never recover that child's trust. When some lawyers renege on their word they worry how much it will cost them to negotiate an acceptable compensation. I knew that if I took John back, the chances were good that he would be unfaithful again.

One day, when we were talking on the phone, John emotionally promised to be a "better husband" if I would reconcile with him. I wish I had asked him to define what he meant by that. I assumed that since he wouldn't admit he'd been having an affair, maybe this was his way of saying he would be faithful in the future. I should have insisted on having it spelled out. I would learn that to him being a "better husband" meant covering his tracks more carefully so I could live in blissful denial about his philandering.

I have to face now that even though my gut instinct kept telling me not to take John back, I was trying to find a way to do just that. Not because I missed John, but because I didn't want to let down Auntie, Hattie, and the doctor. At that time, I thought more about how terrible it would be to disappoint those people who had been like parents to me than I thought about my own happiness.

I also worried about depriving Melodie of her father. The pattern you saw everywhere around you in those days was that in divorces the father drifted further and further from regular contact with his children, most absent fathers eventually losing all contact with the children and even ceasing to contribute to their support. If I had thought about myself first, I would never have gone back into a marriage with a man who I feared was incapable of either fidelity or candor.

But I didn't have the courage then to consider my own needs. Like many women of my generation, I saw my primary responsibilities as lying elsewhere. I wanted to please everyone else first.

I don't know why I never had sense enough to listen to Josie when it came to John. She was the one person I knew who saw through John's charm and didn't care one hoot about how successful he was. More important, in her view of things, what other people might think or say was far down the list of what I should be bothering myself with.

One day, when she and I were having lunch, I mustered up the courage to tell her that I was thinking of getting back together with John. Her eyes widened and she shook her head, "You are sick, girl, if you take that man back. We won't have to worry about anyone else killing you, because I'll handle it myself."

I laughed—even when she was telling me I was a fool, Josie made me laugh. I didn't have the courage to tell her that the reason I was thinking about going back with John was that I didn't want to disappoint everybody else. I had to come up with a better reason for Josie. Looking back, the reason I gave her that day may have been the only halfway decent reason I had. "I'm going to give it my best shot, Josie. I'm going to give it my all. If I don't, I don't know if I could deal with the guilt of wondering if I really tried hard enough to make it work. I might always be second guessing myself. But if I do give it my best and it still doesn't work out, then I can walk away and never look back."

One night not long after that, John and I went out again. We had a good time. It was the end of the evening and he was going through his litany of why we needed to reconcile. "I want to be with you and Melodie. That is my place. We can look for a nicer house in an nicer area. We want the same things, Barbara."

"John," I interrupted. "Let's stop talking about it. Let's just give it a try."

He smiled. I didn't know if it was because he had, after several months of real effort, won another case, or if he was truly happy to be back with me. I knew I had a long way to go to ever feel as happy as I had been when I first saw him in the church the day of our wedding. But I was going to try.

Chapter 8

John and I got back together and—sincerely, I thought—tried to start over. Only later, with the help of Patricia Cochran, formerly Patricia Sikora of the detective's report, was I finally able to unravel the web of lies he began weaving in 1966.

Patty and I had spoken briefly on the phone several times in the mid and late eighties, and we first met at John's mother's funeral in 1991. I had never held a grudge against Patty. Knowing John as I did, I figured that he had probably been telling her whatever lies he needed to get her to go along with him, just as he had been lying to me to get me to stay with the situation—and, God knows, he can be very convincing. We had exchanged a few stories on the phone, but at the beginning of this year—thanks to John—we finally sat down to actually talk and compare notes.

As it turned out, each of us had been outraged by John's attempts in the publicity he was getting from the Simpson trial to paint his past life as wholesome and benign, expecting us both to stand by and passively acquiesce to the false portrait. Neither of us did. In my case, of course, I refused to deny to the press the physical abuse in our marriage and to heed John's request to "just say I'm a wonderful guy."

In Patty's case it was worse. If articles in *People* and other magazines accurately reflect what John told them, he basically denied Patty's existence. The articles mentioned everyone in John's life, including his son by Patty, but not one word about her. Her father and other family members and friends all recall that Patty was surprised at the omission. Both of us decided to continue to speak the truth when the dirt-seeking press asked us for it.

But what infuriated John even more than the fact that we would be telling the truth to the press is that we started to talk with each other, and to tell each other the truth. Once, John called Patty while I was on the other line with her. He warned her, "Don't talk to Barbara! She's a racist. And she hates you. If she calls, just hang up the phone. Don't even speak to her."

Patty thanked him for his concern and got back on the line with me. We laughed as she related his words to me. If he had still been on speaking terms with me, he would probably have called and said, "Don't talk to Patty. She's a racist. And she hates you." He couldn't do so because he had already told me he would never ask me for anything again.

The truth was that John finally found himself powerless over two women he had manipulated for more than three decades, and he was frantically trying to maintain some control over at least one of them. But it wasn't working. We were too busy sharing our stories and untangling the dense web of lies that he had woven around both of our lives.

I'm interrupting the story of my own life with John now to reveal some of what I later learned of his behavior out of my presence when our marriage first started to fall apart—as I pieced it together from incidents related to me by Patty and her daughter, April, in conversations I had with each of them this year and in years past.

Patricia Sikora, a young blonde, blue-eyed legal secretary, was working at the Union Bank in Los Angeles in the bank properties division when she met John. It was just about the time he first went into practice, and he wanted to lease a small office in the Union Bank building on Wilshire. One of the leasing agents in Patty's office felt sorry for the young attorney and revealed to Patty that John had had to use his wife's mink jacket as collateral when he signed the lease. Since Patty was separated from her husband and looking for a divorce lawyer, the agent suggested that she might consider hiring John as her attorney.

She did. John must have been quite taken with the pretty, vulnerable young woman who walked into his office seeking a divorce. Ironically enough, given John's later career, the man Patty was divorcing was a police officer. More important at the time was that Patty had been brought up a devout Catholic and was heartsick that her marriage was ending in divorce—like me, she had been raised to think of marriage as forever.

John took the case and, while representing her, began his seduction. He started sending Patty flowers and gifts—to cheer her up through a difficult time, he said. On his professional letters to her, he began to write endearing little handwritten notes at the bottom of the page. Patty was young and naive then, but she knew that these weren't the kind of things that most lawyers do. She recognized a come-on when she saw one. She also knew that John was married and told him that she didn't want to pursue a relationship with a married man.

But John pleaded his case, eloquently I imagine. He asserted that he was in a very bad marriage. His wife was terrible. She was running around on him. He begged Patty to bear with him while he tried to get free of her.

Their romance began right around the time of the Deadwyler case, when John was experiencing his first taste of fame. He bragged to Patty that women were calling him and telling him they wanted to "eat crackers in his bed." Patty thought this expression—which she had never heard before—rather crude. But she did not think of John as the philandering type. He seemed too much of a gentleman to her. Besides, he had told her that he would always—above all things—tell her the truth. It was like a mantra in their relationship—no matter what, he said, he would always be honest with her.

Their affair soon shifted into high gear. As the "lying" detective accurately reported, John went to Patty's house in Eagle Rock several nights a week. As my "silly," unreliable girlfriends had thought, that was John with another woman on those weekend trips he took to see clients.

Our daughter Melodie went to Patty's house several times during this period, once for the sixth birthday party for April, Patty's daughter. April remembers being introduced to the shy little Melodie as her "sister," even though there was no blood relationship between them. Melodie, a year younger than April, did not know what to make of it all. She must have heard her father calling another woman "dear" and "darling." She must have seen him being affectionate with Patty.

I recall asking Melodie where she had gone with her dad after one of these outings, and that she was unusually quiet. "Oh, just a birthday party," she said. I'm sure she didn't want to risk hurting her mother by talking about what she had really seen.

John took Patty to meet his parents. Here was one of the greatest shocks I endured in searching for the truth. John's parents were introduced to Patty before the day I dumped his clothes at their house. They told her that I was greedy for John's money and was fighting his desire for a divorce every step of the way.

I was stunned to learn they had said this to Patty. These were the same parents who told me that there was no such person as Patricia Sikora, who told me that the detective had lied to me, the same two people who pressured me to reconcile with their son, two people I thought loved me and cared about me as though I were their own child. Yet for their son, these otherwise decent people would clearly lie to anyone, anytime, and about anything.

While John and I were separated, John was around Patty's house so much that Patty's daughter April began to think of him as her stepfather. One time, when she was still about six, she remembers going with her mother and John for a ride in the Cadillac I had financed for him. As they drove around their pre-

98

dominantly white neighborhood, April noticed that everyone was staring at them. She asked why. John answered, "Well, they're staring because your mother is so pretty. And because we have such a nice car."

He didn't mention the real reason why they were staring—that he was black and Patty was white. But in hindsight, April thought that was a good omission. He wanted her not to be overly race conscious. He always stressed to her that black people and white people were the same. She looked up to him and admired him for all that he taught her.

Despite how devoted to her John seemed, Patty was not willing to stay with him unless he showed some evidence of his good intentions. He did. One night at Patty's home, while she was sitting at the kitchen table, he got down on his knees and said, "I want you to be my wife. You're the only woman I love. I'll always love you." Patty was touched by how romantic he was and how sincere he seemed. Her younger sister and her two daughters were there, and they were all so excited that Johnnie had made this commitment to Patty.

Meanwhile, back on Sutro, John and his parents were still pressuring me to enter into a reconciliation. I believed at the time, erroneously of course, that they were so sympathetic to John's position because he was still vehemently denying that he was having an affair, and that they did not yet know of Patty firsthand.

While I was agonizing over whether or not I should go ahead and divorce John, he was apparently trying to figure out what to tell Patty if he won his case with me, and I agreed to reconcile. One day, after John and I agreed to give our marriage another try, he tearfully told Patty that he needed to move back in with me—not because he loved me, not because he had any hope for our marriage—but because I had named Patty as a co-respondent in my divorce action. He had researched the law, he said, and the only way he could save Patty from harassment by me and humiliation in court was to move back in with me for six months. After that, I wouldn't be able to name Patty because I would have "condoned" his affair by letting him move back in. At the end of the six months, he would then be able to file for a divorce.

In big letters he wrote across a newspaper, "Together in 1968." It was a sad time for them, but he promised her that he was only doing this for her—to keep her name from being dragged through the mud in my divorce action—and would be sleeping in a separate room from me.

As I succumbed to John's pressure to reconcile, and Patty sadly listened to his story of why he needed to move back in with me, April remembers one night during that period. She liked John and wanted him to be with them all the time. She was playing with his money clip—he used to let her count his hundred-dollar bills. He was lying on the bed, reading a brief. She asked him, in the inquisitive way a six-year-old would, "When are you and Mom going to get married?" He said, very much like an intended-father would say, "Well, I'm going through a divorce right now, April. Sometimes these things take a long time. But after that, we'll get married."

Shortly thereafter, John moved back in with me. Patty never knew that I hadn't named her in the divorce complaint my attorney filed when he filed the restraining order. The restraining order described John's violence, but the complaint had no more than a standard boilerplate clause about cruelty: " … defendant has treated plaintiff with extreme cruelty and has wrongfully inflicted upon her grievous mental suffering."

Had I actually gone through with a divorce then, I doubt that I would have mentioned Patty. In fact, I would have been surprised to be told that I needed anything more. John's violence toward me seemed more than adequate grounds for a divorce. But an attorney like John, with a propensity for overkill, could easily have imagined that I would throw everything but the kitchen sink at him. At any rate, he had no trouble spinning false tales about the content of my legal papers.

And so Patty and I, both full of hope, each went forward in our lives, we thought in separate directions, but entwined together by the web of lies John continued to weave between us.

Chapter 9

When John returned home, I thought things were different. He took fewer weekend trips. He always called when he was going to be late, and, at first, he came home late less frequently. Now, late didn't mean 1:30 a.m. anymore. It meant 10 or 10:30 p.m. He was still very busy, but we managed to find some time to spend together.

I had always liked basketball and we started going to Laker games together, often with my friend Juanita and her husband. The first time we went, as I watched the game, I couldn't help remembering the detective's report and its description of how John had gone to a game and then to Patricia Sikora's house afterward. Feelings of hurt and humiliation started to rise up in me but I pushed them back down. Learning to forgive wasn't easy, but that was what I had told myself I must do when I decided to reconcile. John was making an effort—or so I thought. To do my part, I had to let the past go.

After John left the city attorney's office and went into private practice, he caught banquet fever. The original "networker," he wanted to be seen at every social event in Los Angeles—dinners, luncheons, brunches, baby christenings, ground-breaking ceremonies, and so on ad nauseam. Unfortunately for me, he always wanted me with him. A lot of times I would rather have been home spending time with Melodie or reading a good book, but I went. I wanted to be a supportive wife.

At one banquet I ran into one of John's ex-girlfriends from

UCLA I hadn't seen in years. By now, I had almost forgotten John's description of the noises she made during orgasm. That evening, instead, I really looked at her for the first time and noticed that she was a very beautiful woman. On the way home that night I remarked to John that she was very pretty and asked him why he hadn't married her. He launched into a long-winded discourse which, I assume was intended as a compliment. "There are women out there, Barbara, who are more beautiful than you are. There are some who are more intelligent." He continued to measure me up against other women on a few more scales—classiness, sexiness, etc.—with at least one of the others exceeding me in each category. Then he came to his big conclusion: "But point for point you are a little bit above all of them."

"Thank you so much," I said sarcastically. He seemed surprised that I wasn't thrilled to hear that I had squeaked by on this point-by-point assessment against such high-level competition.

Since I believed we were really trying to start over, I thought it would be a good idea to get away together more often than we had in the past. Vacations when I was growing up had always been so peaceful, pleasant, and restorative. I hoped time away from home and work would have a similar effect on our marriage. So not long after we reconciled, John, Melodie, and I took a trip to Hawaii. With its shimmering blue waters, sprawling beaches, marvelous hotels, and exotic foods, Hawaii, I thought, would inspire us to rest, luxuriate, and simply enjoy the laid-back islands together. Well, we could have been in our backyard for as much tranquillity as John allowed himself. While Melodie played in the sand and I sat on the beach reading a novel, John nervously fidgeted next to me, flipping through briefs and law journals, composing speeches, rummaging in his briefcase, and writing in his appointment book. He barely noticed the beautiful surroundings. And, of course, he called his office every day.

For the next vacation, I figured that sight-seeing might be a more restful holiday for John and planned a trip to Europe. He probably couldn't read a brief while looking at the Arc de Triumphe or walking through the Louvre, I thought. He didn't even try. But he turned his compulsiveness toward making sure that we saw every possible site. We were out of breath running from one beautiful place to another, with barely a moment to enjoy anything. The only time I recall John ever stopping for a second to enjoy a moment wasn't by his choice. As we were scurrying around London, we lost our way. John stepped up onto a trolley to ask the driver for directions. When he got the information he needed, he turned to get off the trolley. But it had started to move, though very slowly. John decided to jump off. He must have misjudged the speed of the trolley because he ended up sprawled face down on the street. Nothing was hurt but his pride. Melodie and I—and even John—had a good laugh over it. This was the old John. One of the sadder aspects of a good marriage gone bad is that there are so many pleasant memories to go with the nightmares. John always had an excellent sense of humor, and while I don't know exactly when he finally accepted himself as too big a person to be the butt of his own jokes, to be able to laugh at himself, I do know that from that point forward he was a clearly a lesser man.

We then rushed on to Paris where we were booked into a very elegant hotel. At the end of our stay, John decided he wanted to keep one of the luxurious robes that were in the rooms. He told me to put it in my luggage. I protested. I wasn't going to take a robe for him. But at some point when I wasn't looking, John put the robe in my bag anyway. As we were leaving the hotel, the concierge asked to check our bags. He pulled the stolen robe from my bag, looking sternly at me. Embarrassed and angry, I turned and looked sternly at John. Never being one to take responsibility for his actions, John said nothing. The concierge finally closed our bags, and we guiltily left the hotel.

After we had been back together for a while, we started to look for a new home. John was doing quite well in his practice and wanted a bigger house to better reflect his growing stature. By this time he was very much into conspicuous consumption—he had gone far beyond that first Cadillac. I received a mink jacket one year and a beautiful white mink coat another. He also gave me an expensive dinner ring with a big gaudy diamond. I appreciated these gifts, but was never as impressed with material things as John was. I always thought that the best gift he ever gave me was a double album, an anthology of Smokey Robinson hits. It probably cost less than $20, but more than the rings and the furs it was a gift that had meaning because it reflected an understanding on his part about something of value to me personally. It was not something for me to wear to show the world how successful he was. It was something just for me, Barbara—because I loved Smokey Robinson's music.

On April 4, 1968, I was still teaching in the inner city at a mostly black elementary school. That day, the principal came into each of the classrooms to tell the teachers that Martin Luther King, Jr., had been assassinated. I was completely devastated. I continued to admire King, even though several black leaders had started calling for a more confrontational strategy. His loss was incomprehensible to me—a tragic reminder that the anger, hatred, and prejudice in this country hadn't died. Later that night, when John and I talked about the assassination, I felt a deep sense of mourning. John had been saying recently that King had not been as effective as many—like me—thought. That night, he added, "I guess the era of non-violence has stretched as far as it can go." To me, it felt like the hope and optimism of that glorious era was also vanishing. As frustratingly slow as the progress had been, it seemed clear that black Americans and white Americans had been growing closer together under the inspiration of Dr. King and his colleagues in the non-violent Southern Christian Leadership Conference. From the point of his death onward the races in America have grown further and further apart.

Our search for a new house was put on hold when, late in the

105

fall, I discovered I was pregnant. The baby was due in August so I was able to work most of the school year, which was good because it took my mind off that terrible curse—morning sickness. I had it every day of my second pregnancy, just as I had on my first.

It had taken me several months to get back down to my normal weight after Melodie's birth, so I swore I would watch my weight during this pregnancy. And I did gain less—only 49 pounds this time instead of 50. I was crazy about food when pregnant. Some women develop intense cravings for exotic food when pregnant. With me, just drop the word exotic. One day, while getting groceries, I bought a whole watermelon. I came home, sat down at the kitchen table, and ate the whole thing. My doctor had only two words of advice for me during the pregnancy: "Stop eating!"

Juanita and a few other friends gave me another baby shower. Given John's growing stature in the community, fifty women crowded into a small private home would never have been enough. This baby was going to be welcomed into the world in a blaze of glory. More than a hundred women celebrated the coming event in an elegant hotel banquet room. Once again our friends provided everything we needed for the new baby. But this time John was a far cry from the struggling law student he had been when Melodie was born.

For me at least, the key to bringing on labor is playing in a bridge tournament. The night my second child was born I was playing in a tournament with my friend Jennie. We didn't do any better than Myrna and I had done when we played the night before Melodie was born. Perhaps those babies of mine each thought, "Poor thing. She can't play decent bridge. I might as well make my appearance and give her something else to do." Anyway, when I got home about 1 a.m. that night I felt my first labor pain. I called the doctor, who told me that this might be a false labor since the baby was not due for another two weeks, but I should go to the hospital anyway. I hurriedly woke John, packed, and got Melodie up and ready to go to John's mother's. By the time John and I dropped Melodie off and arrived at the hospital, it was about 4:30 a.m. Note that it was John who drove me. This fact becomes important later in the story. I settled in for another long labor.

It wasn't a long labor. It was exceedingly short. The baby was born before my doctor even arrived at the hospital. I had thought that this baby might be a boy because it kicked me all through my pregnancy, and Melodie hadn't done that. John was anticipating a little Johnnie L. Cochran III, I think, but he got a little tiny "Tiffany."

The baby was so small that it frightened me. She was in good health, but I had never seen a baby that tiny—she was only five pounds when we left the hospital. Again, I remembered what had happened to my parent's first child and worried that our new baby wouldn't make it. I asked John to take us by our church on the way home so I could go in and say a prayer for her.

Auntie had come to town for a visit and to help me with the new baby. She was waiting for us when we got home. She saw me waddling up the sidewalk to the house and I overheard her say to whoever was also there, "She couldn't have had that baby yet. Look how big she is." I started on my diet that minute.

Melodie was anxious to have a little companion, but was disappointed when she saw the tiny baby. She asked me, "What kind of baby is that, Mother? Is she ever going to be able to play?" I told her to give the baby a few months. When Tiffany did get old enough to talk and walk, she followed Melodie everywhere and would do whatever Melodie told her to do. Before going off to school, Melodie would tell Tiff, who would be staying behind in the care of a baby-sitter, to watch *All My Children* and report everything that happened when Melodie returned home. One day, when I came home from work, Tiff asked me, "What's an abortion?" I said, "Where did you hear that, honey?" "On *All My Children*," she admitted. "Well, you ask Melodie what it is because she is the one who told you to watch that show." I was curious to see how my nine-year-old would explain abortion to my almost three-year-old. When Tiffany asked Melodie, it wasn't a big deal to Mel. She simply said, "Oh, it's when you are going to have a baby and you don't want the baby. That's all."

John missed a lot of wonderful moments with the children. He got home late so often that he rarely saw them, except on the weekends. I thought that was the price we had to pay for his success. I'd usually see him a few hours each night, and I'd try to keep him up on what was going on in their lives.

Despite some improvements in the marriage and despite the joy of the new baby, these first few years after John and I reconciled could hardly be described as blissful. I started to suspect that either his affair with Patricia Sikora wasn't over or that he was seeing another woman. I didn't know then what I now know, of course, but there were the usual kinds of signs. He was out fairly late many times during the week, and the number of nights per week seemed to creep up as time went on. How many late meetings, dinner meetings, and last-minute briefs to write could one lawyer have, I thought? He also had to go out of town at least once a month. On evenings out with me, he always would get up to go to the bathroom and be gone long enough to make a phone call. People who have had to endure a cheating spouse will likely recognize all these signs, and they all did bother me at the time, though I tried to put them out of my mind.

On Tuesday morning, February 9, 1971, the Sylmar earthquake hit the Los Angeles area. It was a pretty bad trembler that rocked the area early in the morning, well before I usually left for work. John had said he was going to be out of town and hadn't left a number where I could reach him—he usually didn't—which left me alone in the house with two very frightened little children. But within minutes after the quake struck he called to see if we were all right. I was too busy dealing with calming the children and cleaning up the mess to think about the oddity of the timing. But later I thought, if he had been out of town, how would he have heard about the quake so quickly? News travels fast these days, but news of an earthquake travels that fast only if you feel it under your feet.

I tried not to dwell on the suspicious thoughts his behavior engendered—they just wore me down. After all, maybe John really did have to work that many hours. "Yeah, right," I thought. My intuition told me that his absence was due to something other than a heavy work schedule. But it would be many years before I learned the whole truth.

We had gradually fallen into an understanding of sorts, I suppose. Our relationship wasn't wonderful. It wasn't intimate and close the way my parents' had been—I don't think that would have ever been possible with John—but it was fairly comfortable. We didn't fight much because I didn't make many demands on him. I didn't confront him with my suspicions because I didn't want to suffer the indignity of being lied to or jollied. I certainly didn't want to go through those angry confrontations I had gone through before our first separation, and I suppose he didn't either. Maybe there was the specter of the old violence standing in the background. Certainly neither of us wanted to go through that again. And maybe I stayed off the subject because he wasn't throwing his affair up in my face and being so obvious about it the way he had been back in 1966 and 1967.

So I just kept my head buried in the sand and my mouth shut. When I look back now, however, I don't think I should have put up with it. But that's based on who I am these days—and based on what I know now about his activities back then. In those days, I was still a sweet, long-suffering little wife, trying to find the proper role for myself in a painfully disappointing marriage. Like other friends of mine who were going through similar problems with their husbands, I just put up with the prospect that John was likely being unfaithful—as long as he remained at least somewhat discreet about it.

The physical violence was, in my mind, the main reason I had separated from John, and thankfully, that at least hadn't returned to the marriage. But John's aggressive anger, which had not just evaporated, needed new avenues for expression. Even during that relatively civilized period in our marriage, John began to criticize me routinely, in a cruel, taunting manner. My profession was always a favorite target. He would say that he didn't know why I kept my "stupid" job. "Those who can, do; those who can't, teach," he sneered. That wasn't very original, but coming from my husband, it stung. Other times, when I felt sad or upset about something, instead of comforting me, he would ridicule me as a "poor orphan school teacher." This was a man I had always built up with praise and positive feedback, because I understood how important a strong sense of self is to success, and even to every-day happiness. In return I was treated to an onslaught of nasty words intended to cut me down to manageable size by eroding my spirit and self-esteem.

This was the private John I had now come to know. More and more often, I experienced the very mean-spirited person behind the charitable and humor-laced public persona everyone else saw. In contradiction to the bright, charming exterior he donned every morning as he left for work, with me he was consumed with contempt, not only for me, but for other hard-working women trying simply to earn an honest living. When we had our house on Sutro up for sale, a single woman came to look at it. John was home at the time, and he showed her around the house, making small talk with her. When she mentioned she was a nurse, he held forth on how much he respected what she did and how important her profession was to society. She was charmed—she smiled and nodded at him as he talked. Then, as soon as she left, he said, "If she's so bright, why didn't she go on and become a doctor? It's just like you. Those who can't do, teach. She's too dumb to be a doctor, so she became a nurse."

He would have liked me to profess admiration for his instant analysis of the poor woman, but I heard only his duplicity. I said, "John, while you were saying all the wonderful things you said to her, is that really what you were thinking?" He scoffed at me, "Oh, Barbara, don't be so serious."

In the early years of our marriage when I heard John say negative things about people behind their backs, I tried to excuse many of the comments as stemming from his truly creative sense of humor, from his inability to resist a tempting line. Or, I'd think, well, he doesn't hear how he sounds, doesn't mean to be hurtful. But now I understood that he did mean to hurt, that that was the very idea of it all, that when he practiced cutting people with as few words as possible he was sharpening his tongue, maybe for later use in court against a potentially damaging witness. Often the target of his scornful remarks, I came to know firsthand that they were not funny to the target.

At some point after we reconciled, John suggested that we have a "sounding board" on Wednesday nights, where we could talk about things that were bugging us. I figured he suggested the idea to show me that he was really interested in trying to make the marriage work. It was a bomb, lasting about one Wednesday night. We just sat there and watched TV, each of us, I imagine, waiting for the other to start the sounding-board process. He didn't bring up any problems that he had. I was too anxious to use the Wednesday night sounding board to sound off about how I thought he was still having an affair. He would only say, "You're crazy, Barbara." And that would be that. And I didn't know what to say about the other things that bothered me. I didn't have the ability then to describe how hurtful it was when he called me names, or when he didn't appear to respect me or my work. I hadn't seen such insensitivity in my own family. I had seen the opposite. With parents who were considerate and supportive of each other, I had never learned how to deal with an insensitive person, how to communicate my pain without exposing myself to more hurt. I thought it should be obvious. If John had wanted to be kind, I thought, he would have been, without my begging for gentler treatment. If he didn't, then nothing I could say would change him. To the contrary, it might further arm him against me.

The marriage hobbled along. I knew it wasn't great, but I didn't know how to fix it. I hoped something would come along to make it better—maybe time alone would do it.

After Tiff was born, our search for a new house began in earnest again. I had already looked in the View Park and Baldwin Hills areas—nice black middle-class areas of L.A.—and ruled them out. I hadn't grown up in an all-black neighborhood. Used to seeing both black and white faces around, I wanted the same for my daughters. Also, I felt the schools were better in the mixed areas. We were sending Melodie to a private school at the time, but I believed in public education and wanted to be in an area where the children could go to a good public school, and in relative safety. And that kind of environment was more likely to be offered to white children, or to those in integrated schools.

It was just chance that led us to our new house. We were out looking one Sunday and happened to be in the Los Feliz area, a nice hilly section of Los Angeles not too far from downtown. We came to a street and I said, "John, turn up here. I haven't seen this area before." He did. When we got to the top of the hill, there was a house still under construction. We got out and looked around. The house had two stories and five bedrooms. I particularly liked the grounds. A swimming pool with a fence around it was planned for the frontyard, and the backyard was large enough for swings and slides. It was great—we could have a pool and yet not worry that the kids playing in the backyard would wander into it and drown. The house seemed perfect.

The next day, when I called about the house, I reached the contractor. He told me he was building the house for a man and his wife to move into. It wasn't for sale. I asked for the owner's phone number anyway. I really wanted that house and wasn't going to give up that easily.

I called the owner of the house, a Russian immigrant, who lived in a small house down the hill from where he was building. He had purchased several lots in Los Feliz, and this was the first house he was building. As the contractor had said, he and his wife intended to live in it. I told him that since he owned several lots, he could sell this one to us and still build the next one for his wife. Over the next few months, I called him so often that we came to feel as if we knew each other, even though we had never met. As the house neared completion, he finally agreed to meet us there so we could look around some more. One Sunday after church—when we were all dressed to the nines, with Mel, Tiff, and I in matching outfits—we drove up Hobart Drive to meet the owner of our dream house. He was visibly shocked when he saw us. This was an all-white neighborhood at the time, and in all those telephone conversations it had never occurred to him that I was black. He collected himself quickly and was very gracious. He showed us around. Now that it was almost finished, we loved the house even more. We told him we would make him an offer. Who knows, maybe it was the fact that we were black that encouraged him to sell to us. Maybe he had faced discrimination himself as an immigrant and was predisposed to be especially kind to us. Or maybe it was just that our offer was too good to refuse. Whatever the reason, he sold us the house. It was June 1972, and we planned to move in after school was out for the summer.

I was excited. We had found a really beautiful home—with a fabulous view of both the valley and downtown and with plenty of space. John absolutely loved the house. Maybe things would change now, I thought. Maybe John would be home more now that he had this beautiful house to come home to. I was always the optimist. Looking back now, however, I have to admit, I was more often the fool.

Chapter 10

Meanwhile, to return to John's other world—with Patty—he was complaining to her that it was hell living in the same house with me. He wanted her to appreciate the sacrifices he was making to protect her reputation. He painted a picture of me as crazy, irrational, and sometimes even physically violent. On one occasion, Patty noticed a cut on John's shoulder, and he used that opportunity to spin an incredible tale about how he had innocently given Melodie a candy bar called a "Peanut Patty," and I went bonkers. I grabbed a butcher knife and attacked him. It was all he could do to get away from me without more serious injury. When Patty first met me years later, she was surprised that I was so small. From John's stories about me, she imagined someone of Amazon proportions.

When John and Patty went out together, they usually went to movies or plays, but not to social activities. On Saturday afternoons, John would sometimes take Patty to visit the crime scene of one of his cases or to the jail facility where he would meet with a client. He told Patty that he was "shy" and didn't care for large social events. All the banquets and parties he had to go to were just meaningless to him, he said, and he went only because I insisted or because I had already committed us to go with consulting him. What he really loved, he claimed, was the "quiet time" he had with Patty. Of course, she had no way of knowing that in actuality it was always John who insisted that we go to the "meaningless" gatherings. And while John didn't include this one additional factor in what he told Patty, most of the events were the kind of occasions where a black wife at his side was quite useful. Black middle-class people can be fully as conservative as their white counterparts, and walking into one of those staid social functions and introducing his blonde, blue-eyed mistress to all the wives in attendance wouldn't have fostered the kind of impression of respectability John felt he needed to foster, and might even cost him some business. And he certainly wouldn't have been able to cultivate the image of a bona fide defender of racial justice if word got around that he had abandoned his black wife for a white woman as soon as he found some success.

As my children were learning about abortion from the TV soaps, Patty was learning about it more painfully, from her relationship with John. She first discovered she was pregnant in the summer of 1968. John insisted that she get an abortion because the timing wasn't right for them to have a child, since his divorce wasn't final yet. According to John, I was still stalling and fighting the divorce, preventing him from obtaining the freedom he needed to marry Patty. This was before the *Roe v. Wade* decision and abortions weren't easy to get, so John arranged for Patty to go to Japan for the procedure. Of course, he was too busy to go with her, but he sent along a doctor to look out for Patty. This doctor had read about the Deadwyler case in the paper and had called John to offer his medical services to Mrs. Deadwyler, who was pregnant at the time of her husband's murder. John now put the generous doctor to work accompanying his mistress to Tokyo for an abortion. But when they got to their hotel, the Tokyo Prince, John had booked only one room. Patty refused to sleep in the same room with a man she had only just met, even if he was a doctor, and sat up all night in the lobby.

The next day, as Patty lay on the table waiting for her abortion, John called and told her that she wouldn't ever have to do this again. They would be married soon.

Sometime after Patty was back from her traumatic trip to Japan,

John decided that it would be better for them if she sold her house in Eagle Rock and moved to L.A., closer to him. There weren't enough hours in his *Captain's Paradise* life to squander any in travel. He found an apartment for Patty only a couple of blocks from his office. He also decided that they should put the money Patty received from the sale of her house into a joint savings account. He told her he would match the funds in the account; it would be their little nest egg when they got married. Patty told me that he never matched the money in the account. Instead, he withdrew most of the money and then eventually gave the drained account's passbook to their son, Jonathan.

Now that Patty lived so close to John's office, he managed to spend even more time with her and April. After the first passionate stage of their affair, they settled into a more domestic routine. John would stop over at Patty's after he left the office. He'd read, help April with her homework, or watch TV while Patty made dinner for the family. After April was in bed, they might have some intimate time together. Then John would leave and come home to our house. April remembers John being there for "the most important times that a father spends with his children." She recalls, for example, the time she was trying to make sense of what she had heard about the "birds and the bees." She said to her mother, "I understand how the man has the sperm and the woman has the egg. I just don't understand how the man's sperm gets into the woman's egg." Patty decided to let John explain it to April. God knows, he knew a lot about the process. He gently told April, "The man loves the woman and they get close together and they kiss each other." Then he explained the sex act.

April remembers John as a kind and caring father figure to her when she was a young girl. I am glad he was good to her, but as I listened to her stories of their family life together, I grieved for my neglected daughters, particularly my oldest, who was about April's age. Melodie hasn't the same memories of her father being there for her, doesn't remember him teaching her anything about life. She just remembers the large expensive gifts he bought her and put at the end of her bed. It angered me that my children were without a father not because he was working so hard providing for them, not because he was busy fighting for justice for African Americans, but because he was—except for a piece of paper—essentially a bigamist, portioning himself out to two families.

In 1969, Patty found out that I was pregnant. She was at John's office one day, and one of his law partners had an 8×10 photo of me at the lavish baby shower for Tiffany. I looked very pregnant. The attorney didn't realize that Patty didn't know I was expecting a baby. I guess John didn't keep his law partners as up to date on the operative story as he did his parents. It must have been a particularly painful moment for Patty, coming so soon after she had been forced to go to Tokyo for an abortion, and she immediately confronted John about it. Naturally, he came up with an answer, one truly worthy of John. He told her that he hadn't mentioned that I was pregnant because it wasn't his baby. He hadn't slept with me. But I had been sleeping around, and the man who got me pregnant wouldn't marry me. What was a good man to do? Since our divorce wasn't final, he had to give the new baby his name, so Melodie, who was his daughter, and the new baby would have the same last name. But he swore on a stack of bibles that this child was not his.

After hearing Patty recall this classic Johnnie Cochran performance for me—yet another on-the-spot improvisation of truly epic proportions—I figured that if he ever wants to give up the bar he can always find employment writing daytime soap operas or Harlequin novels.

When Tiffany was born, John called Patty and said, "You'll be very proud of me." "Why?" she asked. "Barbara had her baby and her girlfriend took her to the hospital. I wasn't even part of it." A few months after she was born, John brought Tiffany over to Patty's to show Patty that the child didn't look like him. Of course, he also enlisted the help of his parents in this newest hoax. Regarding my pregnancy, Hattie Cochran told Patty, "It's a shame that Barbara pulled that on Johnnie. She knew the marriage was over, and she was just greedy. She wanted more child support from him. That's the only reason she went out and got pregnant." When I learned that Hattie had said this, I once again felt so deeply betrayed. John's tall tales got laughable after a while, but each time I heard of lies by Hattie and the doctor—both of whom I trusted and respected even after I left John—it cut to the quick.

After Patty moved to the apartment near John's office, she discovered she was pregnant for the second time. Again, John told her that the timing wasn't right for them to have a child. She should have another abortion. I, of course, was still to blame since I was fighting the divorce, according to John. Abortion remained illegal, but this time John didn't send Patty out of the country. He drove her to a house in the San Fernando Valley. They went into the den of the house and John paid a man in cash. Patty got on a makeshift operating table for the procedure. She wasn't given anything for the pain and the conditions were hardly sanitary. She remembers the pain well, and the cold, clammy, sweaty feeling she had while the abortion was taking place. Afterwards, she got an infection and John took her to Dr. Herbert Avery for treatment— Herbie was "Georgia Boy," the nice guy I used to ride to UCLA with. I wondered how many other friends of mine knew all about John's life with Patty while I was still kept in the dark.

Sometime after Tiffany was born, John and another lawyer bought a nightclub called the "There" club. John got me to sign a quit claim so the club would be only in his name. He told me, "The other lawyer didn't have any lady's name on it with him." I just went along with what he asked of me, not thinking too much about it at the time. I went to the club once or twice. Little did I know, however, that John had Patty running the club for him. She must have been off-duty the times I was there, but I later learned she managed the club for over a year.

It was now 1971 and John had been pulling off this double life since 1967. There were times when Patty had tried to break off the relationship with him because she felt it wasn't going anywhere. One time, he sat in Patty's house on a large reclining chair he had bought for her and held her on his lap as she told him she thought they should break up, the two of them crying together over the prospect of a future each without the other. April came into the room, saw Patty and John crying, and joined them on the big chair. She didn't want her mother and "Mr. Cockroach," as she teasingly called him, to break up. Needless to say, they didn't.

During this period, I, of course, suspected that John was still having an affair, but would never have guessed the extent of it, that he was actually living a double life. My suspicions of John's whereabouts the night of the February 1971 earthquake, however, were later confirmed by Patty. Although John had told me he would be out of town, he was in fact in town during the earthquake, at Patty's. Patty's daughter April remembers being awakened by the quake and running into her mother's bedroom. John was there. It didn't seem at all unusual to April to see John there. She remembers his being there often, including during another emergency, when a next-door neighbor had a stroke in the middle of the night and John got out of bed and went over to help. As much as I have come to know and like Patty and April, these stories still leave me with pain. While I am happy to hear that April had the attention and presence of so strong a father figure during her childhood, I grieve for the loss my own daughters had to endure to make it possible.

In 1972, as we were moving to the house on Hobart, John told Patty that he had purchased that house as a "neutral" ground where he could see his children. In one more duplicitous tale, he had Patty believing that the new house was purchased to provide a place for John to be with his daughters. In this story, I had threatened that if Patty were ever at his new house, I wouldn't let him have visitation with Melodie and Tiffany. The silver lining, he

119

rejoiced, was that he would be away from me, since I was also off-limits at this new house. If someone had recommended to Patty that she engage the services of a private investigator, as my aunt had done for me, the report would have read that I spent parts of every day on the neutral ground, and all night long.

Shortly after we moved into the house on Hobart, Patty again discovered she was pregnant. This time, John told her he wanted her to have the baby. He was still going through the divorce, and I was fighting him tooth and nail over finances, trying to take everything away from him, he claimed. But he promised Patty that once the divorce was final, they would do the right thing and get married. He wanted everything to be proper. She could count on that, he explained, because he might want to be a judge one day, and they'd have to be legitimate before that opportunity came along.

Patty later told me, "The worst thing, Barbara, was that I believed him. I always believed him." I tried to reassure her that she had not been particularly gullible. "How could you know, Patty? You had never known a person like him. Don't punish yourself over it."

I could have been talking to myself. For far too long I had also believed John. Finally, my eyes started to open.

Chapter 11

When old friends came to visit us in our beautiful new house on the hill, they thought that John and I had really arrived. We had a live-in baby-sitter, new cars every couple of years, a big screen television as soon as they came out, and, of course, a swimming pool. If we weren't yet Hollywood Hills we were at least high L.A.

But John was still not satisfied. He wanted to be a millionaire before he was forty. Whenever possible, he wanted to socialize with politicians, wealthy people, celebrities, people he considered more "important" than our old circle of friends. A well-known singer became his client and we were invited to a party at his house and to dinner with him and his wife a few times. I felt sorry for the man because the only friends he seemed to have were the business-people around him—his accountant, manager, agent and, of course, his attorney. His wife seemed to want to befriend me and once bragged to me, when we were alone, about having had an affair with another famous man behind her husband's back. Given what I had been through with John, my sympathies were with her husband, and from that point forward I felt uncomfortable around both of them. I didn't really want to pursue a closer friendship with her, and so when she sent me an invitation to a shower she was giving, I promptly called and thanked her, but told her I wouldn't be able to make it. When John found out, he had a fit. He told me that this was my chance to socialize with the upper echelon of society. How could I not want to be friends with such people!

I didn't feel the same way about all the "upper-echelon" people we knew. Just like people in other income strata, some are terrific and others aren't. One person I really liked was Julian Dixon, who later became a member of Congress from one of the Los Angeles districts. Whenever we went out with Julian and his wife, we had a good time. He was always down-to-earth and warm.

By the time we moved to Hobart, John was often the center of attention at the parties, dances, or banquets we attended in the black community. A group would always gather around him, asking him about some of his more important cases. The year we moved, John was representing Geronimo Pratt, a member of the Black Panther party who was accused of murdering a woman schoolteacher on a Santa Monica tennis court. It was a very high-profile case. John defended Pratt with a passion and thought he was going to win an acquittal. Late in the trial, John showed the jury a Polaroid photograph that he said had been taken by Pratt's brother in 1968, just after the killing. The prosecution had claimed that the murderer was clean-shaven, and this photo showed the defendant with a mustache and "chin hair." But this evidence that John hoped would get his client off, in fact backfired and helped the prosecution. The D.A. produced a witness from the Polaroid Company who testified that the particular film used in the photo wasn't even manufactured until May of 1969, plenty of time for a clean-shaven person to grow some facial hair. John said that he was "shocked and stunned" by this information. Pratt was convicted and sentenced to life in prison.

While John was holding forth at various events we attended about his more successful cases, I started to notice a number of women staring at me. They would come up and say things like, "Hi, you must be Barbara." I'd hear their names and talk to them a little. I figured that these were women who had silent crushes on my famous husband. It made me a little uncomfortable, but one night it did more than that. A girlfriend of mine came up to me after a conversation with one of these women and said, "That woman you were just talking to? She has been going out with John, you know." I had known no such thing. I tried to put on the jaded act, like, "So what else is new?" but it really hurt. I went home that night in a kind of shock. I realized I had accepted his mistress—as long as he kept the affair rather discreet. But other women, too? How many of the women I had to deal with at all these events was he also running around with? Was that why I seemed to be getting so many strange looks from women I had never met?

It kept happening. A woman would follow me around at a party and then some well-meaning friend would inform me that the woman had been seeing John. It got so I no longer had to hear the reports. I could tell by the woman's body language, how she looked at me or talked to me, whether or not she was sleeping with John.

It was becoming more difficult to go out with John, to smile and act as if everything were fine. It was hard to hold my head up and keep my dignity intact while I wondered how many of the women in this room had been intimate with my husband. When I told Josie about these new revelations, she called John an "alley cat." For once, I thought, Josie had been too kind. The things I felt like calling him then aren't printable.

I guess I had rationalized John's affair with Patty as something he was doing because he was hung-up on white women. John had come along at a time when black male movie stars were first having affairs with white women, something denied to black men in America since the first arrival of the two races on the continent. Whenever we heard of another such case, John remarked about it, as if it were a big deal. And he later on he must have imagined himself a star in his own field and entitled to all the perks of stardom, now a part of that world where the rules were different for black men. I had also heard him talking to a male friend of his about how being with a white woman is a way that a black man can get back at the white man for the generations of humiliation and oppression that the white man has perpetrated against him. He takes from the white man his crowning glory, the white woman, that jewel forever past denied to black men even in their thoughts, under the threat of lynching. So perhaps, I thought, he was just acting out a sort of racial revenge on the white man. When he had enough of it, he'd see he was hurting not all white men who had oppressed black men, but this particular white woman who in choosing him as a lover had shown herself no racist, and himself. Then he might end the affair. These were ridiculous reasons to continue to put up with infidelity, I now know, but I was trying to make a marriage work.

This new realization that John wasn't just hung up on one white woman but was philandering all over town hit me hard. I was quickly losing all respect and admiration for John. After a few of these parties, I quit trying to find time for us to be together. I quit trying to make things work between us. I spent more time with my friends. I became more independent. Looking back, I realize that I started, brick by brick, to build a wall of insulation between John and myself. I started the psychological process of uncoupling.

This time I didn't seriously consider divorce, as I had earlier. Some part of me knew I wanted a real relationship in my life, with love and intimacy, but I was torn between that desire and the influence of the black middle-class world I was a part of. Everybody in the community admired John so much. And they admired us, still the perfect couple, a role model for upwardly mobile young black couples just starting out together. All of my girlfriends—except Josie—told me I was crazy if I even thought about leaving John. If I mentioned that I was unhappy with John, they would say, "He is the father of your children. You don't want to take your children away from their father, do you?" Others friends were even more emphatic: "Don't even think about letting those other ladies run you out of that house. You're the queen of your castle. You just stay put."

So I did. But the emptiness inside of me began to grow like a tumor. I just hadn't yet discovered how deeply malignant it was.

It wasn't long before our lives became more and more separate. I had my world and John had his, our worlds intersecting only at the social events we attended together where his professional needs called for me to be at his side. Otherwise, he was consumed with his work, his white woman and his black women. I had my work, my friends, and my children.

I had transferred schools and was teaching not far from our new house. Melodie was enrolled in the nearby junior high school. Tiffany, when she was old enough, would start school at my new school.

Tiffany was quite an outspoken little hellion at that time. When she was four, about a year after we moved to the big house, I had a group of women up to the house to play cards. A lady came up to me that day and said, "Barbara, you have got to discipline that child. She called me an 'old bitch.'" I said, "Oh, my God. I'm so sorry. Let me go and talk to her." I found Tiff and said, "Please don't call anyone an old bitch. Don't ever do that again." She looked at me, "Mother, she is an old bitch. She looks at me funny and has wrinkles all over her face. What else should I call her?" I shook my head, "You call her by her name. Go over there and apologize to her right now." She did. But it was a constant struggle to civilize that child.

Once, Tiff, Mel, and I were going into the drive-through line at the local McDonald's. I noticed a black man having lunch with a white woman on the patio. Tiff noticed them too. She stuck her head out of the car window and said, "What's that black man doing eating lunch with that white woman! As many black people as there are around here, he could find one of them to eat with!" I just pulled the car out of the line and drove home, too embarrassed to stay. When I got home I told Tiff to never say that again. It didn't occur to me then to suggest that if she had a compulsion to give advice like that she should consider giving it to her father instead of some stranger at McDonald's.

Melodie was very involved in school activities. She played a number of sports and ran for student council president. For her campaign, I helped her think up her slogan, make posters and write her campaign speeches. Her slogan was "Vote for *me*, Melo*die*!" She won. When Melodie graduated from junior high, she was named "Girl of the Year." I was so proud of her. John made it to her award ceremony, but barely. He was several hours late.

One summer after we moved, I took the kids on a short trip. We went to San Diego with my friend Juanita and her children. When we got back, I noticed that our bed, still made up exactly as I had left it, quite clearly hadn't been slept in. I didn't usually confront John, but this time, for some reason, I did. We were both in the bedroom and I blurted out, "Well, look at that. That bed hasn't been slept in since I left." The familiar refrain began immediately. "Oh, Barbara, you're crazy. Where in the world would I be? Of course, I slept in the bed. I learned how to make up a bed from watching you, so it's only natural that I'd make it up just the way you do." Another winner! I let it go. What was there to say?

Despite John's tales to others that I later heard, I never ran around on him. I had no interest in hurting him the way he hurt me. And I respected my marriage vows even if he didn't. But I didn't mind getting a compliment or a bit of attention from other men when we were out. It felt nice to hear something positive after the constant put-downs and negative feedback I got from John. But any momentary good feelings I had were short-lived. If John had heard a man compliment me, on the way home he'd say, "For an ugly woman you certainly get a lot of attention. Out of all my girlfriends, you are the ugliest." I was trying to build a protective wall around me so that these kinds of remarks wouldn't hurt so much, but they still did. Angry at him, I'd think to myself, "You're not the best-looking thing I've ever seen either." But I kept quiet. I usually didn't give him the hard time that he gave me.

One time though, I did. There's a little in perversity in all of us, and this time, out of the blue, I decided to have a little fun with him. At the new house we had two phones—one for the children, which I rarely answered, and one for us. They were both unlisted. One evening a woman called on the children's line and I happened to answer. When I said, "Hello," she inquired, "Who is this?" I asked, "Who do you wish to speak to?"

"Are you Johnnie's wife?" she demanded. I still didn't answer but again asked her to whom she wished to speak.

"Wait a minute," she insisted. "Johnnie said the two of you were separated. Are you his wife? I just want to know if you are separated." Irritated, I finally said, "Well, you know what, I'd suggest that you talk to Johnnie about his marital status. Try him at his office." I wished her a good evening and hung up.

A few minutes later, John called and said, "Did some silly woman call here asking if we were separated or something?" That was when I decided to give him a little zing. I said, "Why would some silly woman call here, John? We have an unlisted number. How could anyone get the number?" He paused a minute and then said, "I'll call you back in a few minutes." I laughed to myself, imagining the conversation John and the "silly" woman were having. A few minutes later he called back, "Barbara, are you sure nobody called up there?" I said, "John, why do you keep asking me that? Who could have called here? You know we have an unlisted number. When you get your facts straight, call me back." It was fun, giving him some of his own medicine. He came home early that night and was still obsessed with the subject. Unable to confirm whether or not his friend had really called, he was at a disadvantage in trying to decide on the story to use with her. "Barbara, did somebody call here, some silly lady? She may have sounded crazy, even to you." I said, "John, you keep asking me that. Who would have our phone number?" Of course, since he would never admit that he gave this girlfriend the number of our second line that he thought I'd never answer, he never found out if his friend was telling him the truth. And I certainly didn't mind causing a little confusion between him and one of his ladies.

We still took trips together—either John and I alone or as a family. I like to travel so I enjoyed them. But I was so cut off from John that I can't say I looked forward to getting away with him anymore. We went to New York often. John liked to shop for suits there and, of course, I like to shop anywhere. I was looking forward to a particular trip because Josie, whose work often brought her to New York, was going to be in the city at the same time. We planned to get together. I gave her the name of the hotel that John and I were staying at, and she promised to call. When it was time to return home, Josie still hadn't called. I was disappointed, but I figured something must have come up with her work.

When I next spoke to Josie, I asked her what happened. She said, "Bobbie, I called that hotel every single day and they told me you were not registered there." I was furious. That night I asked John what happened and after some probing got him to admit that he told the concierge to have us listed as not registered. "Why did you do that?" I erupted, "Josie was trying to call me." "I wanted us to have some time alone and not be bothered by a lot of calls," he said. "You and Josie can get together anytime."

But I knew that wasn't the reason. He just didn't want any of his girlfriends to call—and be in the position of having to explain to them why I answered, or to me why they were calling.

Sometimes, like the time I missed seeing Josie, John's attempts to juggle all of his women made me furious. But by now it was also getting laughable. One Valentine's Day after we had lived up on Hobart for a while, I got the usual flower delivery. John must feel that to keep women around you simply have to send them flowers a few times a year, because he never missed an important day for flowers. This particular day, as I signed for the roses, I happened to notice the signatures above mine. Listed there were the names of two other women I knew he was currently seeing. It stung—momentarily. But by then my protective walls had gotten higher if not thicker. I was at a minimum beyond wasting tears on John and didn't shed one that day. I just put the flowers on the table and enjoyed them. Years later I learned from Patty that on one flower day she was next on the florist's delivery route after mine. She was devastated to see my name right above where she had to sign. Unable to understand why John was sending me, the horrible monster, flowers when we were supposed to be separated, she cried all afternoon.

I was often amazed at how inconsistent John's stories and lies were. Perhaps he was so confident in his power to persuade people that he didn't think he had to tell stories that made sense or tell the same story from month to month. Sometimes he would boast and say that he had a girlfriend with blonde hair and blue eyes. Other times he would lie and say there was no such person as Patty. If I would remind him that he had already told me that he had a white girlfriend, he would say, "It's not true. I must have been talking out of my head." Or more often, he would go on the attack, "You are making that up!" he would bellow. "I never said such a thing."

If I wasn't already "crazy," as John claimed, this kind of bizarre inconsistency should have been enough to drive me there. At a certain point, however, he found himself no longer able to deny Patty's existence to me. He was too proud of the son she had given him.

On a rare day in 1973, John came home early. I could see he was agitated about something. He waited impatiently until the children went to bed and then rushed to tell me what it was.

"I have a son. His mother is Patty." He waited for my reaction. A number of emotions rushed through me like electrical charges. I felt angry, then disgusted, then hurt and humiliated. For a moment, I just looked at him standing there, proud as a peacock. The deed is done, I realized. What good will it do to get upset with him about this? I finally mustered up the dignity to say, "Well, I'm surprised, but not completely shocked. It's just one more thing that you have done to hurt, shame, and humiliate me."

From his reaction to my remark, I think he might have expected me to say "congratulations" instead. Full of indignation, he said, "I didn't do it deliberately to hurt and humiliate you!" I couldn't stand the sight of him another minute. I merely uttered, "So be it," and went off alone. With the door to the bedroom closed, I cried. But even as I was crying, I was angry that he could still make me cry. I hadn't built that wall thick enough, but believe me, after this I would. Having an illegitimate child was, to me, the ultimate expression of contempt a husband can show his wife. The child would stand as living proof to the world of his unfaithfulness, and any attempt by the wife to deny the affair would hurt only the innocent child.

I decided then that I wasn't going to leave him right now, over this. I was first going to deal with the hurt and humiliation, and if I decided to leave at some point, these feelings would be one less thing that I'd have to go through at that time.

A few days later I saw Josie and I told her about John's son with Patty. It felt good to be around Josie when I was in the dumps about something John had done. Her strength and her humor always picked me up again. Josie asked me how I felt. "Just kind of washed out," I said. "I don't know what more he can do to me." But I told Josie that I was going to deal with it. Each time he dealt me a new indignity, it seemed to be making me stronger. The pain went away more quickly each time.

John was so proud of that boy that he couldn't help but mention things about him from time to time. First, of course, he told me his name, Jonathan Cochran. He said that they hadn't named him Johnnie L. Cochran III because he wanted to save that name in case I had a son. I guess that was my cue to say something like, "Gee, thanks, how thoughtful of you to save that name for me." I didn't say anything.

The truth was, however, that after I learned about Jonathan, I had no interest in sex with John anymore. He apparently thought that all of his sexual conquests made him more attractive, but to me they had always been a turn-off. Now his unfaithfulness was a gate with iron bars. When he approached me, I always had a headache or was too tired. He had taken the specialness out of our intimacy and I wasn't about to just take a ticket on a bakery line and await my turn. Also, it occurred to me that if he had a child with another woman, he wasn't using protection when he slept around with her and others, so who knows what diseases he might be sharing with me.

One time when John was gloating about Jonathan I said, "Why are you talking about that child? How do you think I feel about your having an illegitimate son?" Unbelievably, he told me that he would never have married or reconciled with me if I had had an illegitimate child. Talk about a double standard! In his twisted mind, I should accept him for what he did, but he would have shunned me had I behaved as he did. And he didn't think there was anything strange about this.

I met little Jonathan for the first time when the boy was about two or three. It happened completely without warning. We were supposed to go to John's parent's home to take a family picture. John's sisters and brothers would all be there with their families. That morning, John went out for a while and said he'd be back to pick up Mel, Tiff, and me. When he returned, Jonathan was in the car with him. When I saw the little light-skinned boy it was a shock. I knew who he was even though John did not introduce him or say anything about who he was. I recovered my poise, said hello to the boy, and introduced myself. He was a sweet child. It certainly wasn't his fault that my two-timing husband was his father.

Jonathan knew his grandparents and aunts and uncles, of course. But no one in the family said anything to me about the boy that day. Everyone just acted as if everything were fine. When

I got home, my disgust with John was too deep to even say anything. I just piled up more bricks on the wall between us.

The one time our relationship could have ended over Jonathan happened shortly after I learned about the child. John came in one night and announced, "I want Patty to be able to call here on the children's phone." I exploded, "No way. That's out of the question. The minute she calls you on any telephone in this house, this marriage is over." He argued that she might need to reach him about Jonathan. I told him that he would have to figure out another way to communicate with her because I wasn't going to have her calling my house. He knew I was deadly serious this time and backed down. Patty never called. But as I look back on that moment I realize how far he had led me, one small step at a time. Here he was with a mistress and a child by that mistress, a woman and child he had already introduced to his entire family, and I was still trying to preserve the tatters of my dignity by drawing new lines in the sand. As the lawyers say of the negotiation process, "By the inch it's a cinch."

Shortly after I heard about Patty's child, John tried to involve me in yet another drama. He came home one day and said, "You and the girls better not go out for the next couple of days. Some crazy woman said she was going to drive by here and if she saw any of us outside she was going to kill us." That was the day I let him have it.

"I have a job. The kids are in school. We are not going to stay in the house day after day because you can't control your girlfriends. And how is it anyway that you have such poor taste in women that you end up with people who'd threaten to kill your family—they should be threatening you!"

"Oaww, Barbara, you're just crazy," he interrupted.

I'm ashamed to say that it wasn't until years later, after I had spoken with Patty, that I considered a more likely reason for his strange request—that he had told another of his ladies that he lived alone in the Hobart house, and she was threatening to drive by and see for herself.

But whether or not I had been able to see through his frightening story, I knew the man. "For a man whose professional life is so great, your personal life is so tacky it's incredible. Underneath all your ability and your stature, John, you're just an immature child."

"You don't know what you're talking about, Barbara. You're just a poor orphan school teacher. You wouldn't have anything if it weren't for me."

Our arguments always ended on some such note. But this particular fight felt good because I had let off some steam. And there wasn't much that felt good in this marriage anymore. One day not long after this, I got a glimpse of my future with John, and it was even more frightening than the present hell I lived in.

One of John's role models and idols, Bob Robeson, was a black lawyer who was very prominent and well-to-do. He and his wife had matching white Rolls Royces. I vividly remember a dinner party in the Hollywood Hills that the Robesons and the city's other leading black attorneys and their wives attended. John and I were the youngest couple there. After dinner, the lawyers all went into one room to talk and the wives went into another. I overheard the men. They were swapping stories about how they charmed a judge or held a jury spellbound. The wives weren't talking about their work—none of them worked outside the home except me. Most of them were talking about how much they hated their husbands. One of the wives said that her husband had played around on her from the first month they were married. Another lovely woman said that her husband had just bought his mistress a home down the street from where they lived. She complained that she had to see this woman and her husband's child her every time she went somewhere in the neighborhood.

About four of the women there that night had similar stories about their husbands. All were filled with hatred and bitterness, yet none of them considered leaving their husbands. I said to one of them, "What do you plan to do?" She shrugged, "What can I do? I don't have an education. When I married my husband, I was only just out of high school." One of the other women mentioned that she had thought about playing around herself, but said, "What would I get out of it? I'd just become one more person with a bad reputation and probably a disease." These were all very beautiful women, but they didn't find the same thrill in having affairs that their husbands obviously did. As I watched and listened to them, I became frightened. John was already playing around, already had a long-term mistress, already had a child by her. I saw the bitterness, the hatred, and the cynicism in these women's faces. It was like a picture of my future. And a horrible one, indeed.

That night on the way home, John mentioned that I had been unusually quiet at the dinner party. What was wrong with me? I told him that I didn't know the people very well. I continued to be very quiet on the ride home that night, thinking about the high price these women were paying for their elegant homes, their Rolls Royces, their minks and diamonds. I next had to face that I too was already paying that price. The price was my self-respect.

Chapter 12

P atty's daughter April remembers that when she was around twelve her mother started walking around the house in a baggy housecoat. She asked her mother if she was pregnant, but for several months Patty consistently responded, "Oh, don't be silly." April couldn't understand why her mother would want to hide the fact that she was expecting. Perhaps she was embarrassed that she wasn't yet married to John and didn't want her inquisitive daughter asking too many questions and maybe stumbling on to a few answers Patty wasn't yet ready to face.

But by March 6, 1973, Patty couldn't hide things any longer. This was the date that she and John picked for her to have the child. John drove Patty to Cedars-Sinai Hospital in L.A., where they met Dr. Avery, who induced labor. Patty gave birth to a baby boy, and the proud father, John, was the first to hold the baby. Soon, there were visits at the hospital and cards and flowers from all of John's family. I, of course, and my children weren't invited to share in the family excitement. We knew nothing of the great event until that day John finally couldn't hold it in any longer and blurted out to me that he had a son.

John had told Patty that he wasn't putting in a phone at the "neutral ground" Hobart house because he didn't want harassing phone calls from me. But after the baby was born, Patty was concerned that she wouldn't be able to reach John if there were an emergency. I'm sure that's what motivated John to ask me to let

Patty call on our children's line. When I refused, he had to resur-rect his story about not having a phone because I might call him. One time, Jonathan had a 105-degree fever and to reach John Patty had to try the answering service and John's mother. John responded and got over there quickly, so Patty just put up with the humiliating circumstance.

The year after Jonathan's birth, Patty moved to a house in North Hollywood, only a short hop on the freeway from our house on Hobart. Patty gave John $10,000 of her money to put a down payment on the house, which he then purchased in both of their names. Soon after Patty moved into the house, John applied for a loan for about $20,000 from the bank that held the mortgage on the house. This increased the mortgage payment. But Patty never knew how he used that money—nothing was ever done to improve the house.

John was still telling Patty that I was fighting his divorce. By now, this fictional divorce battle, if real, might have qualified as the longest in legal history. Even though he couldn't yet marry Patty, John didn't like the idea of his son having a different last name from his son's mother, so he did the paperwork for Patty to change her name officially to Patricia Cochran, and got his law partner to make the requested court appearance. I never learned if his office billed her.

As Jonathan was growing up, April remembers John coming over on Saturdays. He would pick up the boy, and they'd go to the car wash. In those days you could stay in the car during the wash and little Jonathan loved watching the soap and water being sprayed on the car while he was in it. After that, they'd stop at Toys 'R' Us and John would buy his son an elaborate, expensive toy. I imagine that's when he would also pick up something for the girls in his other family.

But April's predominant image of John during this period was of a man coming through the front door toting bags of food or his dry cleaning or laundry. He was like any father coming home after day's work, stopping just short of calling out, "Lucy, I'm home." The only aspect of this scenario that April found strange was that their home wasn't really his home. She wasn't a prude, and even after she had given up asking when Johnnie and her mother were going to get married, she continued to wonder why they didn't all live together full-time, as other families did, even when the man and woman weren't legally married. If John was indeed separated, marring this almost perfect domestic scene by continuing to maintain a separate residence for himself seemed to make no sense to April. Of course, neither she nor Patty knew that his separate residence was not a bachelor pad, but another happy American family home, fully equipped with wife and two children.

To keep up the ruse that he lived a spare bachelor's existence in the big house on Hobart Drive, and that Patty was the only woman in his life, John would bring his clothes over for her to mend. She would always have to let his pants out. Patty's nickname for John was "Dolly," and April remembers her saying to him often, "Dolly, I can't let these pants out anymore. There's no more fabric." Of course, Patty didn't know that with two families to spend time with, John would often find himself eating two dinners, especially on holiday occasions. I guess that was about the time I remember John going to the doctor and getting diet pills.

Patty always trusted John in those days. Like me, she accepted that he was a busy lawyer, with his meetings to conduct, his banquets to attend, and his clients to see. If he couldn't be with the family for some reason, he always explained it so well. One time, though, her suspicions were aroused. Patty's sister, Kathy, worked at Pacific Bell, and a co-worker said to Kathy one day, "I thought you said your sister was engaged to Johnnie Cochran?" Kathy nodded, "She is." Her co-worker responded, "Well, I saw him in Las Vegas and he was with some lady named Barbara, and he introduced her as his wife." Kathy was upset and reported to Patty what her friend had said. Worried about her sister, Kathy counseled Patty, "I think he's up to something."

Patty immediately called John. She was in tears, "How can you do this? What's going on here? You lied to me." By now, John was skilled at dealing with such bad breaks, and slipped effortlessly into the technique he had used on me countless times— attacking the credibility of the witness. "Patty," he said without missing a beat, "Your sister is jealous of you. She's jealous of us. She would love to break us up. Who are you going to believe—her or me? I'm the one who loves you. I'm always going to tell you the truth."

Patty chose to believe John and even recalls feeling guilty for having doubted him. She thought, well, maybe Kathy is jealous. A true master of deception, John had put on the same kind of virtuoso performance for Patty as he had for numerous juries, for the public and for the other women in his life, including me. Now, as I recall and record all this, I realize that his phenomenal ability to cast a spell that suspends, or even supplants critical judgment, remains intact. If folks on the street are asked about the O. J. trial, a huge proportion will say, "Well, as Johnnie Cochran put it, the glove didn't fit," or "You heard what Cochran said. The L.A.P.D. framed O. J.," fully accepting John's spin and dismissing the mountain of other evidence against his client. Johnnie usually stops on the courthouse steps as he's about to get into his car, and makes a short, pithy statement to the press on a very complicated case, and for so many people that settles it. And it's not just the anonymous general public. Juries very often believe John, because juries are people and people have always believed him. That's why he's as successful as he is. Why should it surprise anyone to hear that Patty and I were so easily taken in by John's seamlessly constructed, sincerity-laden repackaging of reality?

As Jonathan got older, April entered her teenage years and she remembers the kindly father figure she had known as a younger child turning into a stern, angry disciplinarian. She, of course, pulled her share of teenage pranks, like telling her mother she was going to be at one girlfriend's house when she and that girlfriend really stayed at the house of another friend whose parents were out of town. When April got caught, John went into a rage and threatened to send her back to live with her father in Chicago. He told April to sit down and he paced around and cross-examined her. She sat there terrified, fearful that she would be taken away from her mother, whom she loved.

Admittedly, in the above incident April was guilty of a transgression, but John's threats and rages often seemed to her to be more extreme than the offense warranted. And oftentimes his disciplinary pronouncements didn't seem to have a rational basis at all. April remembers a friend of hers from the neighborhood, named Pierre. He wasn't a boyfriend, just a good friend who was black. One day Pierre stopped by while John was there. He sent the boy away and yelled at April, "I don't want that guy coming here anymore." April was surprised and upset. Pierre was a really nice guy, and she thought Johnnie, who had always told April that people were people no matter their color, would say, "Right on, April." But he didn't, and to this day April doesn't understand why.

Of course, back on Hobart Drive, John was so seldom home that the disciplining of our children fell on me alone.

It wasn't until 1995 that April began to understand at least one possible explanation for why John was so full of rage at her during those years. With the birth of their child, John had to weave an even more complex web of deception. Neither he nor Patty wanted young April, in her rebellious teenage years, to pick at the web and maybe reveal to her little half-brother that his parents weren't married. It actually never occurred to April to spill those beans to Jonathan, but secrets have an insidious way of undermining relationships. As it happened, Jonathan didn't find out his parents weren't married until this year—when John's web of deception finally began to give way under the incessant scrutiny of an factoid-frenzied media.

Once, during this period, after John had been yelling at April for something, Patty remembers her bright young daughter making a remark in anger, a remark that Patty never forgot: "There are two things you should warn your children about, mother, the devil and Johnnie Cochran."

Chapter 13

Meanwhile, back at Hobart Drive, the image of those bitter lawyers' wives I had met at the memorable dinner party continued to haunt me. One Saturday not too long after that party, I took the girls to Griffith Park, a large park just down the hill from us. As the children played, I gazed idly at my surroundings and saw a place filled with couples strolling together, often holding hands, some in animated conversation, others quietly enjoying the pleasant day together. A rush of sadness and loneliness came over me at this reminder of was missing in my life. What I had always considered the most intense pleasure of life, that special intimacy that can be born only in love and must be nurtured by long years of trust, was no longer there in my relationship with John.

Around the same time, the husband of a friend of mine left her. She called me frequently and talked about how much she missed him and how lonely she felt. They used to do so many things together, she said, and was heartsick that she could no longer wake up and look forward to those joyful moments with him. I felt deeply for my friend's pain, but I had another thought as I listened to her: if I left John, I wouldn't miss him at all. I'm as lonely now, living with him, as I could possibly be without him.

As this friend was going through the breakup of her marriage, another was planning her wedding and asked Tiffany to be the flower girl. Tiffany was thrilled. I took her to all the rehearsals, but I was hoping John would attend the wedding to see his delighted and beautiful little daughter doing her part in the ceremony. But, as usual, on the Sunday of the wedding, I sat alone in the church. John was "out of town seeing clients." As I watched my child walk down the aisle, I started to cry. I always cry at weddings, but the tears I shed that day were not tears of joy for my friend at the altar but tears of sadness for myself. I thought about all the milestones in our children's lives I had attended alone and asked myself what I was holding on to. If I was going to live my life alone—psychically, spiritually, intimately alone—then perhaps I should stop trying to delude myself into believing that because a blob of almost inert human protoplasm occasionally parked itself in my living room I wasn't already alone. Worse, I knew nothing was likely to change in the years to come. The dream I had once had for our marriage was long since dead. I thought of Martin Luther King's words, "How long must we wait? How long?" and I asked myself, how much longer must I endure this empty marriage for the sake of my daughters, for the sake of appearances? How long must I wait before I can act without exposing myself to later guilt that I had acted too hastily? Sitting in the church that day, I decided I had waited long enough. The time was now. Once I made that decision, a sense of peace came over me.

I noticed the change immediately. In fact, on the very drive home from the wedding I felt as if a huge weight had been lifted from my shoulders. The depression I had been in, probably for years, started to lift. Slowly, in the months that followed, I could feel my natural optimism and enthusiasm for life returning. I knew I had made the right decision—the only decision possible for my emotional and spiritual survival.

I also knew I couldn't leave immediately, without a plan. I felt certain that if I did, he would try to cheat me out of as much of what was due me as he could, including possibly not paying child support for a while to bring me to my knees and make me more tractable in future negotiations. It wasn't just my general antipathy toward John at the moment that made me feel this way. I have talked to some women going through the divorce process, including a few with real grudges against their former mates, who say that at least they never had to worry that their husband, rat that he was, would try to cheat them financially. Then I talked with others who reported their mates would certainly try to do so, in fact might try everything short of hiring a freelance mugger to make sure they were left with nothing. My judgment was that how a person behaves through the divorce process will simply reflect how he behaved toward his mate throughout the marriage. If he put her down throughout the marriage, he'll put her down throughout the divorce; if he was always jealous and accusatory, the start of the divorce will not change him in this regard; and if he was always acquisitive and selfish during the marriage, the spouse better be prepared for some aggressive scrapping over every nickel.

If I was correct in this judgment I was in for a hard time, for during our marriage John had always acted as if the community property concept was part of a communist conspiracy, refusing to share with me property that should have belonged to both of us.

I recall an occasion when John's parents received an insurance settlement and wanted their children and their children's spouses to enjoy some of it. John took the check his parents gave him but never mentioned it to me. His mother called me one day, perhaps curious that I hadn't thanked them for the check, because that's the kind of thing I could be relied upon to do. She asked me what we were going to be doing with the money, and I asked, "What money?" She felt bad that John hadn't shared it with me and sent another check just for me. Here might have been an opportunity for Hattie to exercise some tough love. Instead of covering up for what she knew had been selfish behavior on her son's part, she should have confronted him and demanded that he send his own check to me in the amount Hattie had intended I get.

John, of course, knows the law far better than I. He was always well aware that California is a community property state, which means that all of the assets purchased with funds earned during a marriage legally belong jointly to the husband and the wife, unless one partner agrees in writing to give the other a specific gift. John received several such "gifts" from me, though not because I wanted to give them to him—or really even understood what I was doing. One line I used to hear a lot from people, when they first learned who my husband was, was that in being married to Johnnie L. Cochran at least I had free access to the services of a good lawyer. As long as he considered us on the same team that was an advantage. When he put us on opposite sides it became a dreadful disadvantage.

In one typical instance, he came home and started to flatter me, in a sticky-sweet tone. I wondered what he wanted. A trifle too casually, he soon dropped a document in front of me. He had taken out a loan from the bank, he explained, and needed me to sign the papers. I looked at them and said, "This doesn't look like any loan from the bank." He said, "Oh, yes dear, it is, darling. These are loan papers. If you would just go ahead and sign them, I can get the loan. I'm going to put the money into my legal practice." I read the papers again and I said, "No, John. These are not loan papers. This is something about a house." Then, like Dr. Jekyll into Mr. Hyde, his demeanor changed in an instant. Furious, he shouted, "Oaww, Barbara, you don't even know what you're reading. Just sign it." He stood over me, glaring at me in that menacing way that I had seen many times before. I thought, well, God, I'm not ready to be hit tonight so I better sign this paper.

He was purchasing a home for Patty and this was a document saying that I had no rights to the property. I didn't know at the time that he was using Patty's money for the purchase, though the deed would carry him as co-owner. I guess Patty and I both had the need of independent counsel on this one.

There were other properties John purchased during our marriage—the restaurant/nightclub that Patty managed and three apartment buildings he bought as investments with two other lawyers. In every case, he got me to sign something to put the properties in his name alone. I did not understand what I had done regarding these investments until we were in the middle of the divorce. But as I planned to leave him, long before I came to understand all his machinations with me, I guessed that as duplicitous as he had been with me about money while we were together, he would be even less of a straight shooter when I left. With a man like John, I had to be ready.

My plan was to start saving some money of my own, from my paychecks each month. I wouldn't leave until I had accumulated enough to make it for a while, without any child support or settlement. That way we I would be less vulnerable when he went to his hardball tactics.

One Sunday, after I decided I was going to leave, John, Melodie, Tiffany, and I were sitting in church. I was reminded of the time many years before when John and I were first engaged, and I sat in this same church and had such high hopes for our marriage. Today, I looked at John and could only think about how full of hypocrisy he had become, posing as a devout, God-loving person on Sunday and yet in his daily life behaving in such an irreligious way, deceiving and hurting all those around him. I'd hate to have been privy to his prayers. I suspect they were filled with outlandish explanations for all he had done, contortions of logic, distortions of morality, mutilations of fact, for here was the only smooth-talker I had ever met with the hubris to believe he could con God.

As I listened to the sermon that day, I also realized that I could never turn to a minister to talk about the problems in my marriage. A minister was a man, like John, and he would see things as John did. His own marriage might well be one of convenience, similar to mine. He would probably try to talk me into staying with John because, after all, every man needs the little wife there, providing the comforts of home while he goes out and does what he has to do, or wants to do, or thinks he wants to do. The minister might not have put it that way, but in many churches, including my own, there definitely seemed to be one set of rules for women to follow and another, more lenient set for men.

In January 1977, I went on the last trip I ever took with John, to New York and Washington, D.C. John had contributed to Jimmy Carter's presidential campaign and we were invited to the inauguration. John's parents went with us. By this time I had so little interest in spending time with John that what would ordinarily have been an exciting time for me wasn't. Before the inaugural ball, we went to New York for a few days for my birthday. Then I came home and John went on to Jamaica to meet "the guys." Yeah sure, I thought. It had gotten so I didn't believe anything he said. Years later I learned from Patty that he met her then for a little vacation. But once I had made my decision to leave, I didn't much what he did. If he had brought home a harem, I probably would have let each woman choose the bedroom she wanted and put him on a schedule, moving him from one bedroom to another to see his ladies.

I knew the end was near, though I wasn't sure yet when it would be. I was just marking time in the marriage, building up my savings account. John must have sensed something had changed because he got more testy and things got more tense between us as the year wore on. That summer, I had planned to take the children to Europe, without John. When I was supposed to purchase the tickets—for no apparent reason other than to try to exercise control—he refused to give me the money for them. I didn't bother to fight with him about it. I just went to my credit union, got the money, and we left.

Mel, Tiff, and I had a wonderful three weeks in London, Paris, Madrid, and the Costa del Sol. I felt so relieved to be away from John. As I traveled with the children, I thought about how John was only their financial support—I was their emotional and moral support, nursemaid, troubleshooter, and buffer against the world.

They will be fine without him and probably will see him more with regular visitation than they see him now. My life would be fine without him, too. I was grateful that he hadn't completely destroyed my self-esteem. My job, my children, and my friends had helped me preserve some small kernel of dignity. In a healthy atmosphere, away from John's constant verbal and psychic abuse, I could nurture myself and find again the lively, energetic, hopeful person I was before I married John. I had taken too much for too long. I decided I couldn't wait any longer. It was time to stop living a lie. I had a real life to live and was eager to get started in it.

After I got back from Europe, I decided to tell John that I was unhappy and that it might be best for us to go our separate ways. I wasn't at all surprised by his reaction. He told me I was ungrateful. He had provided me with so much and yet I was still unhappy. When I didn't take that bait and argue with him, he started on another tack, "You better not leave. You're too ugly to even get a date!" he bellowed, adding, "You'll never find another man like me." I laughed and asked, "Can I get that in writing?" He said he didn't think that was funny. I said it wasn't a joke. I was dead serious. Since his pronouncement that I'd never get a date didn't appear to change my mind, John continued, relying on his primary means of control, "You'll never last six months without my money. You don't know when you're well off and you'll rue the day that you even thought about leaving."

The main reason for a woman to stay in a marriage, in John's mind, was money. Money and the gifts it bought were the sum and substance of what a woman had a right to expect from a man. From his point of view, this need of women for money and security were tools to be used in their control, to keep them in their place. He seemed to understand this way back at the beginning, for many times during our marriage he pressured me to quit my job. He suggested that I deserved an easier life, occasionally adding that I could be more useful, as well as happier, if I became active in one of those women's groups that supported professional men in their careers, like the "Wives of the Bench and Bar Club." My full-time job kept me from fulfilling this duty to him.

As I look back with the advantage of hindsight, I realize that the smartest thing I ever did was keep my job. As he seemed to know that my job was an impediment to his gaining full control over me, I seemed to know as instinctively—or perhaps it was our first separation that taught me—that without my own income I would be completely trapped. The diamonds, minks, and new cars I received were not spontaneous gifts of love. In return for them, I was expected to play my assigned role, to be there for him, a smiling ornament at his side for public events and privately to be willing to put up with verbal abuse and infidelity without complaint. An important element of this gift-giving was the implicit understanding that it could be diminished or even withdrawn at any moment, if ever my behavior failed to meet his expectations.

How is it that we can be put in such fear of losing things we never wanted in the first place? I'll leave the definitive answer to this one for the psychologists to figure out, but I would guess that in many cases where these things become so important for us it is because over time we have allowed them to crowd out of our lives the things that used to be important to us.

At last I understood that this bargain, this implicit contract he had made with me, benefited him, not me. It gave him what he wanted and me what he wanted me to have. Under its terms, I was slowly dying inside, turning bitter and cynical like the other wives of successful men I met who were trapped in the same type of giver/grateful, bestower/beholden contract.

Here's where I was luckier than most women in our circle. By refusing to give up my job, which provided me with my own career and my own colleagues, and by keeping my old friendships intact, despite John's many attempts to drive wedges between my friends and me, I avoided having the furs and jewels and cars replace those things of real value in my life. As a consequence, when I had to face giving up the all accouterments of John's success, I didn't have to face giving up everything of value to me.

Sometime before I decided to leave John, my Auntie Annette moved from El Paso back to L.A. I invited her to stay with us until she found a place of her own. She came, with Doris and Doris's children, and stayed for a few weeks. The house was crowded, but bearable. John, in his typical generous manner, said to me, "I buy a big house and you fill it up with all your relatives."

Now that Auntie was back in town, it was inevitable that I would talk to her about what I was feeling. I told her that I was really unhappy with John and was thinking of leaving. Her advice, unfortunately, was similar to her advice when I separated from him the first time, "Stay right there and tough it out, Barbara. I wouldn't let all those women run me off. And I certainly wouldn't leave my home. I'd make him get out and I'd keep the home." At that time there weren't too many well-to-do black men out there, and Auntie, like many other women, felt that if you left one middle-class black man, you had to go right out and find another one in order to maintain your standard of living.

It's a quirk of human nature that when people give advice to loved ones they feel an obligation to package that advice so that it protects the loved one's financial interests above all others. Maybe it's a commentary on our times that we seem to put so much store in material possessions. I listened to Auntie, but this time I was prepared to disappoint her. This time I was strong enough to make my own decision about my marriage, rather than bend to the pressure of others.

Each of us has to rely for guidance on what had been important in our own lives. If we have always been happiest when we had money, then go for the money. If we have had money and have still been unhappy, it makes sense to look at the price we've paid for money. And make no mistake about it. Money can be the most expensive commodity we bargain for. It has a high acquisition cost and a higher maintenance cost.

I decided I didn't want to stay in our big house and go through the agony of trying to get John out. He loved the house, and I expected I'd probably have trouble getting him out, but more important, it didn't hold that many pleasant memories for me. The simplest thing, I felt, was for me to leave and find something I could afford. Part of my new life, I figured, was being in a new place.

After the school year started, I began to look for a place. I wanted something not too far from where we lived now, because of the children's schools and my work, which were both located nearby. I saw an ad, for an upstairs of a duplex with three bedrooms and two baths. After school one day I looked at it. It was spacious and filled with light, just perfect. I gave the landlord a check then and there.

As I left the apartment, I thought, my God, I have really done it. It was scary. Now I had to figure out how I was going to move us out without John trying to stop me. The first thing I did was tell Josie what I had done. She looked at me, compassion in her eyes, "Bobbie, you have fought a good fight." I burst into tears, not because the marriage was over—I had already cried enough over that—but because Josie really understood what I had been through all these years. She knew how painful it had been for me, and in that simple, supportive statement she acknowledged it. I was so grateful for her understanding, and also for the help she offered.

Over the next few weeks, Josie went with me to the apartment several times. When she first saw it she smiled and said, "This is nice. You'll be happy here." She introduced me to a guy who put up draperies and curtains throughout the place. It wasn't long before it started to look really cute. Once or twice, as I watched my dear friend Josie taking her time to help me out, it crossed my mind that if I had listened to her earlier, my life might have been very different. In our relationship, Josie was definitely the teacher and I the student. And I had been a very slow learner.

At some point, while I was planning my move, John brought up the subject of separating again. "I don't think you really are, but if you are planning on leaving, let me know because we can at least sit down and talk about dividing the furniture." I nodded. I had no intention of talking to him about dividing anything until I had my own lawyer.

I set a date for the move on a Friday in October, before a weekend when John said he was going out of town. I knew I had to get a lawyer before I moved because, with John's history of violence, I might need another restraining order. At the last bridge club I had up at the Hobart house, I asked one friend to stay behind a minute to talk to me. She was older and had gone through a rough divorce herself. She also seemed to be someone perceptive enough to see through John's exterior charm and to comprehend what he might be like to deal with. I thought she might know a lawyer who would be willing to go to battle with him. She was very supportive and gave me the name of a lawyer, Don Rosen.

In Don Rosen's office, I was a very different woman than I was a decade earlier when I went to Stanley Poster after John hit me. This time, I knew exactly what I wanted—a divorce. Don Rosen agreed to communicate with John or his attorney after I moved, so I wouldn't have to deal with him for a while. He understood exactly where I was emotionally and what I needed from him. He said, "You divorced this man years ago. You're just coming to me for the paperwork."

A few days before the move, I told the live-in housekeeper that the children and I were moving out. I wanted her to go with us and assured her that she would still have her job. She agreed to move with us and to keep quiet about it until the day we moved. But I didn't tell the children then, not only because I didn't want them to tell John, but because I couldn't risk their trying to change my mind. I also knew the move wasn't going to be painless for them, and wanted to give them as many worry-free days as possible.

Finally the day arrived. As John left that morning I smiled and waved good-bye to him, thinking with pleasure that he would never be coming home to me again. The movers arrived a short time later. They began packing and carrying furniture out to their truck. I had decided to take just about everything. John made a lot more money than I did and could afford to buy new furniture. He might have to postpone buying a new girlfriend her Cochran Club diamond or a mink, but other than that, he'd be okay.

Everything was going smoothly until the phone rang. It was one of the secretaries from John's office. Her voice was anxious, "Barbara, the security patrol in your neighborhood called here and told Johnnie that either he was being robbed or you were moving out."

As she described how John got really incensed, I imagined the veins bulging out of his forehead. He wanted to rush home, she said, but the other attorneys in the office convinced him not to, saying, "Man, don't go over there. You're just going to hurt somebody or get hurt yourself." She said the coast was clear for a while, because the other two lawyers had taken John out to lunch, but she added, worried for me, "You better hurry. I don't know how long they'll be able to keep him away."

Grateful for the warning, I told the movers to hurry up. They were real troupers and worked as fast as they could. I was tense, checking the window every time I heard a car go by. But John didn't appear. Finally the truck was loaded, and we drove away. If John came home that day, it wasn't until after we were gone. All that was left in the house was John's desk, the large screen TV that he was so proud of, and all the wall photos with him in them. And in the rush, I also left some clothes of mine that I had decided to give away.

When we got to the new place, Josie was there, ready to help. As the movers brought things in, she and I fixed up each of the girl's rooms so it would feel like home for them. Josie stayed on when I left to pick Tiffany up from school. Tiff was in the third grade that year. I told her that we were going to be living in a new place. We weren't going to be living with John for now. She was a little sad, but when she came in and saw her room with everything the way it was before, she settled into watching TV and playing. Melodie, who was fifteen then, had some after-school activities to attend that day. I had given her the address in the morning and told her to take the bus to this address after school. When she walked in and saw all of our things she said, "Wow, what's going on?" I told her we were going to live here for now. After she got over her initial astonishment at how quickly everything had happened, she was okay. The easiest way to get them to accept the situation, I thought, was to make the new place a nice, comfortable home for them.

Josie had dinner with us that first night. After she left and the girls were in bed, I sat in my new living room and looked around. I knew there would still be some grieving ahead of me. I knew it might get difficult financially. But I was so glad I had finally left. I smiled to myself, recalling the words of the old spiritual Martin Luther King, Jr., used to end his great "I Have a Dream" speech: "Free at last! Thank God almighty, free at last!"

PART III

Life After Johnnie

Chapter 14

A few days after I moved out, I got a call from Auntie. "Barbara," she said, "Johnnie's called me. And he really would like to talk to you. He'd like to take you out for dinner and see if there's some way you might be able to work things out and get back together. Can I give him your phone number?"

There was that old feeling in the pit of my stomach again—from here on the pressure would be building, just as it had the first time I tried to separate ten years before. But this time I was older and wiser. I told her John had already brought me enough pain, and I no longer wanted any part of him. She didn't argue but asked me to think about it. Over the next few weeks, she called a few more times, telling me each time that John had called her again, and wondering if I had reconsidered. I hadn't.

John did not become the successful lawyer he is without tenacity, and eventually he got my phone number and called himself. "Barbara, you didn't have to leave like that, you know," he began. "We could have talked about it." Yeah, I thought, we could have had a Wednesday night "sounding board" about it, and everything would have gone smoothly, as long as the bottom line came out John's way.

"There is nothing to talk about, John," I responded. "I don't have any talk left in me."

He then demonstrated anew why it was useless to talk to him. "Well, I really think we should get together and try to talk about

157

patching things up," he said. I was aghast. Hadn't he listened to a word I had been saying over the past several months. This call started off with his saying that I hadn't had to leave the way I did, in other words talking about the method of our separating, but even before he got me face to face he had shifted to talking about patching things up, something I had already clearly communicated I had no interest in doing. I once knew of someone who was a super salesman, despite the fact that he was hard of hearing. At first I wondered how he could be so successful at a line of work that required such good communication skills. I later learned how wrong I was, that in fact this person became successful only after he developed the technique of turning down his hearing aid during a difficult sales pitch, sparing him the distraction of his customer's objections. Whenever I tried to negotiate with John I was reminded of this story, for that was one of John's favorite ploys— to play hearing-impaired salesman. What he didn't hear, or pretended not to have heard, he didn't have to deal with. My best recourse, I figured, was to treat him the way I would one of those persistent telephone solicitors. In a firm but pleasant voice I said, "No, I don't think so. But thanks for calling." He promised to call again.

Several years later, John mentioned to one of my daughters that I was the only woman in the world who had ever walked off and left him. That was apparently a crushing blow to his ego. He wanted to be the one who called the shots, decided whom he would be with, for how long, and when it would be declared over. By the time I left, however, I didn't care who was recorded as having left whom. I just wanted out. I had talked to him about separating long before I left and had given him the opportunity to make it a mutual separation. He played hard-of-hearing, imagining, I suppose, that if I got no concession from him on that tack, I'd simply give up. But he was miscalculating. He apparently hadn't noticed that the sweet, passive little woman he married was no more. Seventeen years with him had toughened me up.

True to his promise, John called again and again in the next few months, always with the same request—to have dinner and talk about a reconciliation. On one occasion, he shifted tactics, now telling me, "I want you to know that I'm willing to go to counseling and all of that. I just want us to try and put our marriage back together." When I heard that I almost laughed, but instead I told him, "You can go ahead and go to counseling if you want, but I'm not giving this marriage one more minute. I have given it too many years, and I'm tired. Really tired. I'm tired of you. I'm tired of your infidelity. And I'm tired of your dishonesty with me."

For perhaps the first time since I'd known John, he didn't go into his usual denial routine. In his most earnest voice, he said, "I'll change. I'll be faithful." But it was much too late for him to be making promises to me. Too much damage had been done. I said, "You know what, John? I don't believe you. And I'm not going to put myself back into a situation where it matters to me whether you're telling me the truth or not."

These kinds of calls went on for about six months. John would also call frequently to ask me to be the hostess for a party he was having or to attend a banquet with him for the mayor or some other important person. This was the "By the inch it's a cinch" strategy. But I had finally learned to say no. And I just kept saying it, to every single request. I wasn't going to be in that position again of saying yes and then having to explain why I was saying no to a next request that represented such a small, incremental erosion of my position. No more drawing ever-retreating lines in the sand.

Even though I was sure I had done the right thing, there were moments after I left that I felt very sad. A dream I had cherished for so long had finally died. Actually, it had died a long time ago, but I had just buried it and now had to go through the grieving stage. Though I had given it my all and done my best, I had a feeling of failure about the death of that dream. Intellectually, I understood that it takes two people to make a marriage work, two people with the same vision for the marriage. Emotionally, it was another matter, and I could not help taking on some of the guilt.

This gnawing sense of failure undermined my faith in marriage to the point that I thought I could never again be happy in that kind of relationship. Sadly, I vowed to myself that I would never marry again.

A couple of months after I left John, Auntie called, but this time it wasn't to plead John's case again. She had been having some pains in her abdominal area. They were so bad that she had stayed in bed for several days. This didn't sound like Auntie who was always on the go. She thought it was time to see a doctor, and wanted me to take her. Cedars-Sinai Hospital, where we went, did a battery of tests. When they were finished, the doctor told her he wanted her to come back the next day and be prepared to stay overnight in the hospital. That frightened me, but Auntie was cheerful, putting up a brave front.

When I brought her back, she had another set of tests. After all the results were in, the doctor gave us the bad news—she had cancer of the liver. I was devastated. I tried to act strong for her as we left the hospital, but I couldn't hold out for long. When we got to the car, I burst into tears. She was only fifty-nine, and so young looking, so energetic. I just couldn't believe this was happening. As we sat there, she amazed me when she said, "Don't be sad for me, Barbara. You'll see, I'm going to be one of those people who go around and tell everybody how I beat cancer." Auntie, in her indomitable way, seemed to deal with even cancer as just another challenge. Do your best to beat it, and then have no regrets.

For the next three months, Auntie had chemotherapy every week. I was the only one she wanted with her at the hospital, so I'd pick her up after school every Wednesday and drive her to her treatments. As terrible as they were, they seemed to be helping. She was doing so well that I thought, maybe she's right, maybe she will be one of those people who tell others about their victory over cancer.

John had hired Bob Robeson as his attorney, and they settled the visitation schedule quickly—every Wednesday, as well as every other weekend—but kept filing continuances about the rest of it. It was exactly as I'd expected, and I'm glad I had saved some money. I had no idea when the divorce would be final.

On Wednesday nights, Tiffany and Melody spent the night at the Hobart house with their father. From what I heard back, there apparently was a different girlfriend playing hostess for them every Wednesday night. After a while, Melodie started to tell John that she needed to stay home and do her homework, but Tiff always went.

It wasn't long before John started to skip taking the kids on his weekends, or only take them for part of the weekend. After they were with him one Saturday, Melodie came home in tears. When I asked her what happened, she said that when they were at Patty's house earlier in the day, she had asked Patty in the course of conversation if she was coming to the party that night. Patty said, "What party?" And Melodie said, "Well, Dad's having a party tonight." Later, just before he dropped her off at home, John took Melodie aside and called her a "stupid bitch" for mentioning the party to Patty. The poor child hadn't known that one of John's other girlfriends was throwing the party, and John hadn't wanted Patty to know about it.

I was so furious at him I was ready to explode. Involving your own daughter in your deceptions can only teach her duplicity. Melodie did not see the value in becoming skilled at such deception and was both confused and hurt. First I did my best to comfort her, assuring her that she hadn't done anything wrong. Then, rather than pick up the telephone and call John right then, I waited until the party started, preferring to speak to John's father, who would more quickly understand why John's behavior was so destructive of his daughter. There was also the fact that I did not expect to be able to get anywhere with John directly. Knowing him as I did, I was sure he would have told me that he didn't owe

Melodie an apology. After all, he would say, he had done better than apologize. He had given her some money or some expensive gift last week.

When I reached John's father, I told him exactly what had happened. "Oh, I am so sorry that John would do that," he said in a quiet voice.

"Doctor," I said, "I will never be satisfied unless you can get John to apologize to her. She deserves an apology now. No matter what kind of party he's having up there, this is more important. He needs to apologize to his daughter tonight."

Not too long afterwards, the phone rang. It was John calling to apologize to Melodie. She forgave him, but like some of the verbal abuse he heaped one me, it was a hard one for her to forget.

The first time I saw John after I moved out was in court a month or so later. I could tell by the smug look on his face when our eyes met that he was getting ready to screw me royally. But when he got up and pranced around before the judge, poor-mouthing in one of his most outrageously expensive suits, I thought he had gone too far. If the judge looked at the suit carefully, he could have figured out that John's income was much higher than he was admitting to. Later, I did an imitation of John for Josie and she fell out on the floor laughing as I mimicked his slimy-sweet tone. "Your honor, this woman spends over two hundred dollars a month getting her clothes cleaned, and oh, your honor, this woman is so extravagant about everything." John barely let the judge get a word in. My lawyer responded to a few of the more absurd contentions, but I told him he didn't have to bother to answer most of the others. As I sat there and listened to John, I just wondered, how could I have wasted almost eighteen years with this man?

The judge finally ordered John to pay me $300 a month in child support, $150 per child. Considering John's income and mine, it was a very low payment, even in 1977. My lawyer was upset, but at that point I knew I could support myself and my children and didn't care to try to maintain anything close to the old standard of living.

John did even better for himself when it came to the marital settlement. John and his lawyer kept filing more continuances, preventing any decision at all. My lawyer informed me that John had properties he was trying to avoid sharing with me in the settlement. He wanted my permission to go to the mat with John, but I was too sick of the whole thing. I told Don Rosen, "Just leave it alone. I don't want the hassle." Much later, I did have some regrets that I didn't let my lawyer fight harder, not only because I deserved a fairer settlement, but because I realized that once more I had allowed John to have his way with me by wearing me down.

Word that I had left John and was walking away from our marriage with so little for myself spread quickly among the many people in the community who knew of us. I started to hear through the grapevine some of what was being said: "She is the biggest fool in the world for leaving him," or, "For a man who makes that kind of money and who is going to make even more money, I would sit up there on top of the hill and never budge."

That mindset, that a person's worth or a marriage's worth could only be calculated by the amount of money the husband earned, was pervasive in the world I traveled in. But these kinds of statements no longer had the same influence on me they once had. I was beginning to recover my self-respect and that was something that no amount of money could have ever purchased.

Josie stopped by a lot after the divorce, much more often than when I was still living with John. She always seemed to need to borrow a dress, but I'm sure she also wanted to check up to see how I was doing. She tried to get me to go out and invited me to attend some of the plays sponsored by the cultural center she worked for. "Come on," she'd say, "After all those years wasting your time with the superficial, material part of life, you need to be introduced to more meaningful things." At times I would go, and

I'd enjoy myself, but after the weekly parade of banquets I had endured for so many years, I was just glad for the most part not to have to go to any social functions at all—meaningful or not.

And if I felt little like going out, I felt even less like dating. Auntie, despite her own much more serious problem, was worried about me. She scolded me, "You can't sit home and let life pass you by, Barbara. You're still so cute and perky. You know there's someone out there for you." But I told her that the most important thing for me now were my daughters. This was a tough period of adjustment for them, particularly for Tiff, who was younger and seemed more upset about the divorce than her sister. I wanted them to be absolutely sure that I was there for them.

But Auntie had never been one to give up easily. One Friday morning when I was about to leave for school, Auntie called. She said, "Put on something cute today, because I want you to come on by here right after work. Don't go home." "All right," I said. "But I think I'm okay the way I'm dressed." But she was insistent, "No, go back and put on something really cute. I'll see you later." Well, I didn't change my clothes because to do so would have made me late for school.

That evening when I stopped by Auntie's, I saw that there was a man there I had never met. I had the impulse to turn around and leave—but I couldn't do that to Auntie. She introduced us and then took me aside. Looking with disapproval at my workday outfit, she told me she was fixing dinner and said, "Go on back home and make yourself look nice." I felt like seventeen again, the age I was when I moved in with Auntie, but I followed her orders.

When I came back, we had a great evening. The guy was laid back and a lot of fun. He asked me out, but I told him I needed more time. I was certainly glad I had met him though, and over the next few years I did go out with him many times. Tiff didn't like him—or anybody I went out with those first few years. I think she was still hoping I would get back with her father. I tried to let her know, as gently as I could, that that was a lost cause.

At the end of the first school year after I left John, my school changed to a year-around school, which meant I wouldn't have my summers off. I was in a panic. I knew I needed that vacation time, not only for myself, but also so I could spend more time with Tiff and Mel. I contacted a friend of mine, a principal at another school, and she turned out to be a life-saver, arranging for my transfer to her school—which was not going on a year-around schedule—and also helping me get a position in the teacher training program. In the years that followed, I worked two Saturdays a month teaching teachers, which helped financially when things got tight after Melodie was in college.

That summer, I had hoped to go on a cruise with Auntie and the girls. But as the summer approached, it was clear that Auntie was getting sick again and wouldn't be able to enjoy it. When her symptoms started to return, she told me, "Don't take me to the hospital until I can't make it at home anymore." One Sunday morning, she called me and I rushed over. She was very weak. I called the ambulance and they took her to the hospital.

She died that Monday evening, June 19, 1978, a devastating loss for me. Auntie had been such a strong, loving presence in my life, especially after my father died over twenty-five years ago, when she stepped in and took over the job of raising me. She had saved my life then and now she was gone.

Along with my leaving John, Auntie's death was one more event that forced on me a heightened sense of maturity. Now I had no one to turn to for advice other than my peers, and no one to rely on but myself. That acceptance, though it saddened me deeply, also made me stronger.

The divorce proceedings were still going on at a snail's pace. Every time we had a date to go back into court, there would be another continuance. By now, even though I knew he was busily figuring out how to keep money and assets from me, John and I had developed a new relationship—an outsider watching us interact might have seen us as friends. Of course, it was so much easier to deal with him now that I wasn't married to him. And once it was finally clear to him that I wasn't going to reconcile, he started treating me less like an ungrateful underling and more like an old friend, one who didn't share his values or interests but with whom he had a certain history. He still called to invite me to parties or events, though I usually didn't attend. After a while, he started to talk to me about his girlfriends. One time he mentioned how two women, both dating him, got into a shouting match over him at a Beverly Hills beauty salon. Barely concealing his pride, he revealed to me what he had heard of the cat fight, verbal blow by verbal blow. He just couldn't believe that these women would carry on like that about a man in public. I said, "John, believe me, I am more shocked than you are that two grown women would carry on like that about you."

For many months, I never went back to the Hobart house, but finally there was some event—I think it was a party for his parent's anniversary—that got me up there. When I rang the doorbell, John answered and said, "Welcome home, Barbara." That was enough to make me feel like turning around and leaving, but I stayed. It was wonderful. I breezed in, said hello to people and socialized, knowing that anytime I wanted, I could leave and go home. The best part, however, was not having to feel the humiliation I used to feel, wondering which of the women at the party had had affairs with John and which he was still trolling on his line. I also knew that the next few days would not be spent listening to his bizarre explanations for some revealing remark one or another woman had made at the party. Having confirmed for myself how free I was of that sordid existence was re-exhilarating. When I got home that night I thanked God—leaving John was the best thing I had ever done for myself.

For Melodie's high school graduation, I threw a party at my house. All of Melodie's family and friends were there, and John came with one of his long-time regulars. I had known about her for many years—I think for a while we shared a florist's delivery route. She was terribly anxious and kept following me around all night, asking if there was anything she could do. I hoped she didn't feel as if she might have been the cause of our divorce—after all, who knew what tall tales John was telling her? For whatever reason, she was driving me so crazy that I finally found John and said to him, "John, go be with your girlfriend and calm her down."

Melodie had been accepted at UCLA and was due to start classes in the fall. I was enormously proud of her. The three of us took a second trip to Europe to celebrate her graduation. But when we returned, I was not quite prepared for the emotional shock of my firstborn going off to college. Melodie planned to live in the dorm, and I'll never forget the Saturday we moved her there. I helped her put away her clothes and organize her little half of the dorm room. Once we had it all fixed up, there was no reason for me to hang around, so I got ready to leave. I told her to promise to call me. She walked me to the car and we hugged. I was fine—so far. Then I started driving away and it hit me. I began to cry so much that I had to pull over.

I had such trouble shaking myself out of it that I called Josie, who immediately told me to come on over. When I got there I was still crying. "Oh, Josie, I said, "What am I going to do? How am I going to make it?" She shook her head at me, "You sure didn't act this broken up when you left John." Naturally, within a few minutes she had me laughing at myself. Josie loved my daughters and had been watching them grow. She warned me, "I hate to tell you, Bobbie, but this is just the beginning of the end. Once they get to college, they know everything. Unfortunately, that's all they know. But for sure it's going to be many years before you're wise, old Mommie again." I guess she was right. There is nothing that forces you to face where you are in your life more than seeing your children, who only a short time ago had been helpless and dependent in your arms, turn right before your eyes into full-fledged adults with their own independent lives.

In December 1977, a couple of months after I left John, L.A.'s then district attorney John Van de Kamp invited him to join the D.A.'s office in the number three position, and John accepted. Even before I left, John had been talking to me about taking this possibility. Even though it meant a drop in income, he knew that to become even more successful than he already was, he needed to cultivate relationships with people in the local, mostly white, power structure. This position would give him the opportunity to learn from the inside how the D.A.'s office worked. And, when he went back to his private practice, he would have both valuable new experience and new contacts that would aid him as a defense lawyer.

When John left his job at the D.A.'s office a few years later, they threw a big going-away party for him. He really wanted me to come to the party, but I was tired and turned him down on this one. With Melodie in college, she needed money for clothes, gas, food, books, and repairs on Patty's old car that John had given her. John paid Melodie's tuition and room and board only, which at UCLA at that time wasn't that much, and I paid everything else. To get by, I had been both working an extra job and also going to school two nights a week, for which I was paid a stipend by the school system. So the night of his party I just wanted to stay home, take a bath and relax, which I did. But after the party, when he brought the children home, John came in to see me. He proudly displayed the plaques and awards that they had given him and explained why he had gotten each of them. As I listened to him, I realized that this man, who had done so much to hurt me, for some reason still wanted my approval. I complimented him, "That's wonderful, John. You must have done a great job. I'm real proud of you." When he went home that night, I'm sure there was somebody there waiting for him, but to me he just seemed like a very lonely man. A man who so much wanted to be loved, but who didn't have the first clue how to give love.

Don't take this wrong. I didn't harbor any thought that John was changing at his core, not for a minute. For every new glimpse of humanity I saw in him, there were a dozen more examples of the same old duplicitous tricks he had been pulling for so many years. One time during this period, I needed to talk to John because of some emergency with one of the children. He wasn't at home, so I called Patty. It was the first time I had ever spoken to her. I explained who I was and why I needed to reach him. She was very nice and told me that Johnnie had gone to a stag party in Las Vegas with some of his friends. I asked her if he had told her what hotel he was staying at and she said no. He was supposed to be calling to tell her the name of the hotel.

After I hung up, I thought, this sure sounds familiar. On a hunch, I looked up the number of John's condominium in Palm Springs and decided to check there. A woman answered, whom I vaguely remembered being associated with John. I asked her if she had seen John. "Hold on a moment," she said. When John came to the phone and heard my voice, he seemed confused. "Barbara, oh my God, what do you want?" he asked me.

I said, "I just want to know when they renamed Palm Springs Las Vegas? If I could just get that straight for the Chamber of Commerce ..."

"Oaww, Barbara," he said, "How did you find me?"

I told him I had called Patty, and she said he was in Las Vegas, but I put two and two together with what I knew of him and came up with Palm Springs. All he could do was laugh. It was really funny to him. But to me it was just one more incident that reaffirmed for me that my decision to leave him was the only possible decision for me. I thought, here is a man who will probably be doing this kind of thing for as long as he has the strength to drag himself into some woman's bed.

When the divorce was final in 1981, John mentioned to me that I could go down to the courthouse and pick up the papers if I wanted to. I didn't bother. In the final judgment, John had to pay me a settlement for my share of the former marital residence, the home on Hobart Drive, and of course child support and the girls' college tuition. He took several years to pay off the settlement.

Had I read those divorce papers then, as I have now, I would have learned much earlier the extent to which I had been had. Of course, had I listened to the advice of my lawyer, I would have understood that community property laws are not some sort of female bonanza. They are based on the altogether reasonable concept that a marriage is a fifty-fifty partnership, and that all wealth accumulated over the course of the partnership should be shared equally by the partners. My lawyer tried to get me to fight because in virtually every case where John had created exceptions to this legal concept, he had used his superiority in legal training to distort what he was doing—never explaining that I was, in effect, giving him a "gift" of my share of what should have been jointly ours. Also, the few times I was reluctant to sign one of the quit claim deeds, he bullied me until I was sufficiently intimidated to sign it. The divorce papers reveal that John even tried to keep me from getting my half of the house. He claimed that most of the money for the down payment was earned by him during the short time we were separated the first time. Fortunately, the judge didn't buy that story.

As a matter of fact. I later learned that the state legislature had women in my situation particularly in mind when writing California's community property law. My case was almost the classic case of the woman who makes substantial early-year sacrifices toward helping her husband establish himself in a lucrative career, and who deserves to be compensated fairly if the marriage later ends in divorce. I married John while he was still in law school and did whatever was necessary to make it possible for him to complete his training and to help him become established as a lawyer, never counting who was making the greater financial contribution at the time. Then, rather than look for the most profitable line of work for myself, I used what time I had left over from a teacher's work schedule to maintain a home for John and our daughters, leaving him free to make his own career decisions based not on child-care worries but solely on the basis of what would generate the most income. By any standard of fairness the woman who helped make him the successful lawyer he is today should have shared more equally in the rewards made possible by those early sacrifices. But like many women who just want to get out of abusive situations without more trouble, I kept telling my lawyer not to fight him and just do what was necessary to get through the divorce process as quickly as possible.

Now, many years later, I realize I was wrong to have been so passive. John tried to cheat me because he has a faulty sense of fairness, but was able to do so in part because I let him, allowing him to intimidate me after we separated, just as I had during our marriage. I began to think that perhaps my share of the responsibility in the failure of our marriage might lie in this very reluctance to confront him during our marriage. I let his behavior go unchallenged many times and by accepting it, in a sense, condoned it. In effect, my going along with his behavior allowed him to rationalize that his behavior couldn't have been too bad or I would not have tolerated it. If I had challenged him more often, he would have been forced to reconsider the nature of the relationship or we would have divorced years earlier, either alternative better than what happened.

I often wonder whether it was the lingering legacy of violence in those early years that prevented me from standing up for myself—I know this is precisely the way it is with many women who have experienced physical abuse in their marriages—but in my case it was also the deception with which John filled our life. I often bit my tongue because I thought, Why bother to raise an issue when he will only lie about it?

Another significant factor in my tolerating so much was the almost overwhelming acceptance in the middle-class community in which we lived of unfaithfulness and spousal abusive on the part of men, particularly wealthy men. Too many women of that time used to say, "If he wants his freedom, let him pay for it." The irony is that women who adopt this posture willingly surrender their own freedom solely to deny their husband his. Although I didn't leave for exactly that reason, I was greatly influenced by my peers. For every Josie who said to me, "I'll kill you myself if you go back to that guy," there were dozens of other women who said, "You're crazy if you leave him." By going along with the role of the wife as tenacious terrier, with her teeth locked onto her husband's income, I allowed John's behavior to continue to victimize me—his verbal and emotional abuse, his dishonesty as well as his long-term, flagrant infidelity, in my view a form of emotional abuse. For many more years than I should have, I stayed in this marriage, keeping me from experiencing the happiness that I craved and that I now know I deserved.

It wasn't till many years later, long after the divorce was final, after years of living alone with the children, after years of living within a protective shell, that I finally let someone into my life who gave me the gift that John had never given me—simple, sweet love and happiness.

Chapter 15

Patty and I had shared John for ten years, though not exclusively of course—there were always other women he lavished his attention on as well. When I left John, I deprived him of the pretext he had been using for not marrying Patty, and it wasn't too much longer before she too began to lose faith in him. But with John's child to raise, Patty kept trying to find ways to make it work, as I had back in 1967.

Perhaps the most amazing thing that has come out of the bizarre life that John put us both through is the friendship that Patty and I have developed. She joked once that, "We are like two sisters who grew up in the same dysfunctional family and who, in order to heal, need to talk over what we have each gone through." Indeed, we are two women caught in one man's web of deceit who have turned to each other to help ourselves recover. As part of that process we have shared many stories since our first tentative attempts at communication and then friendship we started in the mid-eighties. Some of the stories have been painful to hear and others have been hilarious.

When I moved out of the Hobart house in a rush, I left behind some clothes I no longer wanted. One day not long after I left, John showed up at Patty's house and opened the trunk of his car. There were all my clothes, now freshly dry-cleaned. He said to Patty, "Here, I've brought you some clothes. They're all your size." Grateful for the clothes, Patty didn't ask where they came from. She never would have dreamed that she was being asked to wear the clothes of her enemy, the evil tyrant Barbara who for so long had kept John and her from living a straight life by spitefully denying him a divorce. It was only after she and I began to talk, and she found out that I had not only lived at Hobart, but had not left John until 1977, that we figured out that those were my clothes John gave her.

Shortly after I left, John brought Patty up to the house. He said it was a very special night. "I am kicking Barbara out of my life," he proclaimed. He had a change of heart, he said, and would no longer let me intimidate him into not allowing Patty at his house. When Patty arrived for the first time, she was surprised that there was no furniture. She noticed the indentations in the carpet where the furniture had been, and John told her that for the occasion, he had gotten rid of all his old furniture. He wanted her to help him redecorate the house. Then he opened a bottle of champagne and they looked out over the city and toasted his new freedom—from me.

For Patty, this was a special moment. She thought her life was about to change. But it didn't. If anything, it got worse.

When John took the job at the district attorney's office and suffered a cut in income, he asked Patty to help out by going back to work. She did, and John arranged to take her paychecks directly—presumably so he could pay her bills and, I imagine, some of his too.

April remembers some tense times during this period, her late teens. She came home one day and found John and Patty arguing. Patty started to cry, and when John slapped her and started pushing her, April screamed at him, "Stop it! Don't touch my mother!" John turned on April in a rage and told her to get out of the house. She ran out crying, got into her car, and started speeding down the street. Fortunately, the police pulled her over fairly quickly. She was still crying and the officer asked her what happened. She told him, "My parents are fighting and I just had to get out."

It wasn't long after that incident that April moved out of the house. John, I'm sure, was still worried that she might reveal to Jonathan the fact that his parents were never married. For her eighteenth birthday, John gave April her first and last month's rent on an apartment. April was in a lot of emotional pain at that age. She didn't understand why Johnnie had not encouraged her to go to college—she had always gotten good grades in school. And she didn't understand why her mother was often angry or upset with her. It wasn't until years later that she and Patty could honestly talk about this period and heal the rift between them. They discovered that John had been firestarting—saying one thing to April and another to Patty, all designed to turn mother against daughter and daughter against mother, apparently hoping that ensuing friction would push April to move out. Once again John's proclivity for manipulating people in underhanded ways had caused suffering for an innocent victim.

Patty was beginning to feel more and more unhappy in her relationship with John, but at the same time she felt trapped. She had her young son to think about. She was now working full-time. She was still living in a separate residence from John. And she was beginning to suspect that he was involved with other women. On occasions when she was at the Hobart house, she would often hear John whispering on the phone. Once she had the nerve to ask, "Who was that?" He told her it was a client. But she wasn't satisfied, "Why were you talking so quietly?" He turned on her angrily, "Why don't you trust me? I'm always honest with you." The main problem in their relationship, he explained, was that she didn't trust him.

It is interesting that John picked the most logical inference to be drawn from his behavior—that there was a good reason not to trust him—and made Patty feel guilty for thinking that. The irony is that both Patty and I afforded him a degree of trust far beyond what his behavior warranted. It reminds me of the old movie *Gaslight*, in which the husband, played by Claude Rains, used similar methods to drive his wife, played by Ingrid Bergman, crazy. He made noises on the roof, and when she expressed concern and fear, he tried to calm her down by assuring her that there had been no noises, that no one else had heard them. John had tried to "gaslight" both Patty and me for many years. After seventeen years as co-victims of his mind games, we are both lucky to be out of his grasp with our sanity still intact.

Fortunately for Patty, John finally got himself caught a few times and she gradually came to understand that it was not paranoia on her part that caused her to mistrust him but only a healthy skepticism based on the man's track record. One time he took Patty out for a nice dinner to celebrate some special occasion. Over coffee, he said he wanted her to come over to his house afterwards, but she said she was tired and preferred to go home. He seemed disappointed, but dropped her off at home. After being there for a few minutes she had a change of heart. She felt a bit lonely and imagined him sitting alone at home, just as lonely and maybe rejected as well. On a whim, she decided to pick up a bottle of champagne and go over and surprise him. At John's house, she rang the bell and was stunned when a woman answered the door. It was the same woman who had been driving me crazy at Melodie's graduation party.

I don't know what wild story John told Patty to explain away that evening, but she soon found out that John had taken this same woman with him on a trip to New York.

This was the first time that Patty found John guilty of lying "beyond a reasonable doubt," and she must have felt similar to the way I felt when I first saw his affair with Patty set down in detail in a detective's report. John never really explained to Patty the role the other woman played in his life, never said he was sorry or that he had made a mistake. He just sort of glossed over the incident as a misunderstanding all around. The problem had not been caused by his having another woman at this house but by Patty's showing up and barging in on him unannounced. I'm sure he expected Patty to put up with his unfaithfulness the same way he had managed to get me to put up with his long-term relationship with Patty.

Patty took action now, just as I had in 1967. She went to a lawyer to have her house put in her own name. And she insisted that if they were going to continue to see each other, she and John must go to counseling. He agreed. The counselor saw Patty and John separately and together, in a number of sessions. Afterwards, the therapist was very blunt with Patty. He told her that John had a narcissistic personality. He uses people and isn't capable of true intimacy because he sees everyone only in terms of how they can enhance his own personal aspirations. When I heard this I thought, what a smart guy. This counselor had figured out in a few sessions what it had taken me seventeen years to learn. The counselor was also hard on Patty. He told her she was like a doormat to John.

She listened attentively to what the therapist had to say but still felt trapped. She didn't want to give up trying to work things out—sound familiar?—for the sake of Jonathan. Patty tried to forgive John, but for her, as for me, things were never the same after the first confirmed betrayal. It wasn't until Patty and I began talking, of course, that she learned about the many times John had betrayed her before this, as I learned about certain of his betrayals of me I had not known about.

Patty was a born romantic and had saved many mementos and nostalgic items from her life with John—letters, cards, photos, airline tickets from trips together. There were many moments in their relationship that were special and sacred for her, where the things said were important and had special meaning. But to John, she was to realize, things said during these moments were just words he used to engender in her certain attitudes toward him, or at times, induce certain behavior.

One of these romantic moments was staged for Patty during this difficult time for her. John picked her up one day and took her to his jeweler. He had ordered a special diamond for her and it had finally come in. She picked out the mounting and he picked out his wedding bands. They were finally going to get married, he promised. And this ring was hers, no matter what happened, he said, a permanent symbol of his enduring love for her. Even though she was still having trouble trusting him, Patty was caught up in the moment. In the days that followed, she tried hard to forgive and forget John's betrayals. No man was perfect and John certainly seemed to be trying to make it up to her. There are many ways to say you're sorry.

It's hard to imagine how Patty felt when John came over not long after that and told her that he had gotten married. April happened to be there and was stunned. How do you deal with the fact that the man with whom you have lived as daughter and father, who had helped raise you for almost as far back as you can remember, and was finally engaged to marry your mother, was now married, but to someone else. What did he call the situation he was in? Who had he married, she wondered? Who would marry him, knowing that he had this longtime common-law family? Had he gone back to his ex-wife?

Of course, it wasn't his fault, John told Patty. Dale, his new wife, had entrapped him. He just kind of got caught up in things and before he knew it, he was married. He realized now that he had made a terrible mistake and would start making plans to undo it at the appropriate moment. He said that it wasn't really a sexual thing with Dale—she was kind of asexual. He told Patty that he had no doubt in his heart that he and Patty would one day be together again. He made it known that he still wanted to sleep with her. At first, Patty was much too upset. But John kept coming by, acting as though nothing had changed, as though his getting married were nothing more than a new wrinkle in their relationship. One night she decided to go to bed with him. She wanted to see if he would cheat on his new wife the same way he had cheated on her. He would and did.

Soon afterward, Patty learned that Dale was wearing the diamond ring that John had ordered for her—the ring he had said would be hers no matter what.

After spending about the same amount of time involved with John as I had, Patty finally started to see the kind of man he really was, and would likely always be. It was about that time that she and I first began to share our stories. At one point she said to me, "My daughter April went through the boxes of memorabilia I kept from my relationship with Johnnie and said to me, 'Mother, you had everything with this man, everything except marriage,'" I blurted out, "Lucky you!"

We laughed then as we have many times since, grateful for the healing power of laughter and truth.

Chapter 16

My daughters and I lived happily for many years in that first apartment we moved into when I left John. When the landlord put the building up for sale, however, I figured it was time to move. John's sister, Jean, had become a real estate agent, and she helped us look for a house. Melodie was still in college, so Tiff and I looked for months without finding what we wanted. Finally, we noticed a two-family duplex for sale in a nice area. The place seemed perfect, but buying it would mean I'd have to become a landlord myself. I didn't much like that idea, but I knew that sometimes you have to take a risk. Jean wheeled and dealed and got me the property for a price we thought was fair.

We liked our new place, and I eventually adjusted to being a landlord. It was just one more job during those busy years of trying to earn enough money to support a college student and a teenager. We lived upstairs and rented the downstairs, and the building didn't cause me too many problems. It was, however, an older building, and to add the number of electrical outlets we wanted it needed new wiring. I hired a man I knew from church to come in and do the work, and he did it over a several months, dropping in to do a little here and there. One day he brought another man with him. I was dressed in my robe and slippers when I answered the door, trying to relax after a hard day. The electrician introduced me to the man with him, who I thought looked a little like Smokey Robinson. His name was David Berry. My maiden name was Berry, so we spoke for a moment to see if we might be related, but found out we weren't. Then the two of them went to work and I went back to what I was doing.

Every time the electrician came by over the next few months, he brought his friend David with him. He was a nice guy—very polite and very much a gentleman—and we'd always spend a moment or two chatting. After they left one day, Tiff said, "Mother, that guy who looks like Smokey Robinson likes you." I told Tiff, no he doesn't. It's just his way to be polite. Besides, I told her, I wasn't interested in anybody right then.

By now, I had gone out with several men, spending pleasant evenings in settings that might be called dates. John, of course, had told me that I was too ugly to get a date, so I was pleased when various men asked me out, if for no other reason than to prove John wrong. The fact that I was Johnnie Cochran's ex-wife intimidated a few men, one of whom told a mutual friend, "There's no way I can follow that act." Little did he know that the last thing I wanted was anything that might vaguely resemble a reprise of John's "act."

One man I went out, a high school principal, was very nice, but he was too much like a father figure, always taking care of everything for me. After all those years spent as the object of a man's constant manipulation, I couldn't stand to have anybody tell me what to do, under any circumstances, so I knew it couldn't work out. Another man I went out with talked of marriage, but I told him, nothing personal, that I wasn't ready to think of remarriage to anybody. It was true. I didn't want to be another man's vassal, to risk degradation and humiliation again. I knew that all men didn't look at marriage that way, but from what I had seen of John and so many of his friends, the odds didn't look all that good, and I wasn't ready to take the chance.

So when Dave asked me out, I told him I was too busy. He said okay, but kept coming by with his friend, always passing a few pleasant minutes with me before getting to work. One day, he asked if I'd mind if he gave me a call once in a while, and I gave him my telephone number. He started calling, just to chat. Tiff, of course, noticed that he was calling, and that I generally had a smile on my face as I spoke to him. She told John that a guy who looked like Smokey Robinson liked me. His response was, "Well, he probably looks like Smokey Robinson's grandfather."

By this time, John had called to tell me he had met a young lady from the South that he liked. She was called "the pearl of the Southeast," he said. I told him that I was very happy for him.

One weekend when John's new fiancée was in town, I called her on the phone to welcome her to L.A. My children had met her and thought she was very nice, and since she might one day be their stepmother, I wanted to establish my own cordial relationship with her. I found her open and friendly, and as time went on grew to like her very much. Like Patty, Dale always treated my daughters very well, and any parent who has ever been in my position will understand how grateful I was to both of them.

After David Berry had been telephoning for a while, and we had become friends on the phone, Tiff sat me down and had a talk with me. She said, "Mother, I don't know if you realize it but pretty soon I'll be graduating from high school. When I go to college, I intend to live on campus just like Melodie, and you'll be left here by yourself. You better start noticing some of these men. I don't want you to be left here alone." I was touched by how sweet—and how wise—my young daughter was. Fully aware of how much I counted on her to keep me company, she wanted me taken care of, so she could go off and live her life with a clear conscience.

The next time David asked me to go out with him, I took Tiff's advice and accepted. He later told me he couldn't recall ever having to work so hard to get a woman to go out with him, and I believed him, because he was as cute as he was patient. On our first date he took me to a romantic restaurant tucked into the canyon in Beverly Glen, the Four Oaks. We sat by a fireplace and had a wonderful time. On another occasion he took me to Chasens, a spot favored by the Hollywood crowd. He showed me where Alfred Hitchcock used to sit and told me that Elizabeth Taylor loved the chili. Still a fan of Liz, I had to try that chili.

Little by little, Dave and I became really good friends. That Christmas he was invited for dinner with his sisters. We always played cards at my house after dinner on Christmas, and since Dave liked to play cards, I invited him to come over after dinner. When the doorbell rang, I opened it and will always remember the picture of him standing there. I looked at him as if I were seeing him for the first time. If John's thing was expensive suits, Dave went for the shoes. That day he had on a beautiful pair of gray alligator shoes. My eyes traveled up from those shoes to his gray suit, his powder blue shirt and blue tie. He had on the full threads, as he used to say. I looked at that man and I thought, I'm feeling something I haven't felt in a long time. That night I began to think of him as someone with whom I might consider spending the rest of my life.

The fact that Dave hadn't pressured me or rushed me, giving me time to get to know him, helped break down those walls I had built up while I was with John. Gradually, I was able to trust another person again, and my fear of marriage finally dissolved in Dave's kind and sensitive presence. When he proposed on Valentine's Day in 1986, I accepted. We planned to marry in July.

During the six or seven months we dated, I had discovered that Dave was the opposite of John in many ways. Perhaps most important among these ways was that Dave was always honest and straightforward, not just with me but in his dealings with all people, and he expected the same from others. After we married, Dave accidentally encountered a minister at the church we attended on his way out of town with a woman the minister wasn't married to. From the circumstances, it was clear that the minister was having an extramarital affair, and it bothered Dave. He told me that he couldn't listen to that man preach anymore, that he was turned off by hypocrisy. We soon started going to a different church. In the same situation, John probably would have given the minister a broad wink, patted him on the back, and given him some tips on how to deny everything if his wife ever got suspicious.

I had Josie over to dinner to meet Dave before we married, and this time around Josie approved fully of my choice. "Girl, you've found yourself a good one this time," she said with a smile that showed how happy she was for me. As always, Josie was right.

Juanita and Myrna threw a fabulous lingerie shower for me, inviting about fifty friends. Everyone was more mature by now so I didn't get teased about sex—just about how many showers some of them had attended for me. One old friend said jokingly, "My God, you certainly have gotten your share of gifts from me." It was true.

We held the wedding in a hotel. It was a beautiful wedding, though much less elaborate than my wedding to John. Melodie and Tiffany were my bridesmaids and they also gave me away. John's parents and his sisters all came with their husbands, but John, even though he was invited, didn't. I later heard that he said, "I'm not yet ready to watch Barbara walk down the aisle with another man." Apparently he hadn't been ready to have me watch him walk down the aisle either, because I hadn't been invited to his wedding to Dale, which had taken place the year before.

My wedding night with Dave was a new experience for me, full of a sensitivity and sweetness that I had not known before. In the time ahead, I was to find the same tenderness in all aspects of our marriage.

We went to London on our honeymoon. Dave had never been there and we tried to see as much as we could in the eight days we were going to be there. One night, after a long day and then a play, Dave became ill when we got back to our hotel—he felt a shortness of breath. I called the hotel doctor, who gave him a sedative and recommended he see a doctor when he returned home. He was a little tired from the heavy sedative, but was okay the rest of the trip. Still, it frightened me enormously. When we got home, Juanita gave me the name of a doctor and Dave went to see him. The doctor put him on a blood thinner, and after that he seemed fine.

Life with Dave was the opposite of life with John. We enjoyed just being together and doing simple things like taking a walk. We liked to travel and one year took a wonderful romantic trip to Italy, where Dave had also never been. Brought up a Catholic, he was fascinated by the Vatican and the Sistine Chapel, along with all the other art work inspired by religious themes. We looked at paintings and walked and talked, enjoying the sights and each other; he especially loved riding around Venice in a gondola. At home we were just as compatible. Dave liked to play cards and enjoyed movies. And, unlike John, he didn't mind seeing a good love story with me.

The only difference between us was that I had my dozens of friends and Dave was somewhat of a loner. One day, he asked if we could talk about something. We sat down and he told me that he couldn't understand why I was always running off doing things with my girlfriends. He felt lonely when I was gone. I appreciated his feelings, but at the same time I didn't believe in being the type of woman who just drops all of her friends when she gets married. The solution we came up with was to bring him along to some of my activities. He was hesitant at first. We'd go somewhere and all my friends would run up and hug him and kiss him on the cheek. He later told me, "Barbara, I'm not used to all these ladies coming in and kissing me on the cheek. Do they have to kiss me?" I said, "Dave, they just love you." That seemed to give him a new perspective. "Well, that's okay then," he said.

There were so few things that were problems in our marriage, but when there were problems, Dave and I could always talk about them and work them out. I was bothered by his tendency to procrastinate. He would say he would take care of something

185

around the house, but then time would go by and he wouldn't have done it. If I told him I was getting upset about it, however, he'd respond, "Oh, you're right. I'm sorry. I'll get to that tomorrow." And he would.

This was a new experience for me. With Dave, I was able to express myself, even my displeasure, and so was he. With John I had learned to bite my tongue, fearing that if I spoke out about something that was bothering me, he'd strike back, his retaliation most often taking the form of a cruel put-down. With Dave it was so different. I thought, my God, he understands my needs and respects them. It felt so wonderful.

Josie was as busy as ever with her job. Still, we always made time to see each other, and I'm sure she was grateful I was no longer grieving over my dead marriage with John. She called one day to make plans to have lunch the next week. We talked on the phone for over an hour that day. She told me how stressful her job was and complained a little about her high blood pressure. She said the doctors were having difficulty getting her medication right and she wasn't feeling as good as she should. "I don't know how I got this high blood pressure," she said, "because I'm even thinner than you." I had an idea how she got it. "Are you still smoking?" I asked. She mumbled, "Well, what can I tell you, Bobbie? I've got to have my cigarettes. I've just got to have them." I begged her to try to cut back. Like many smokers, she just changed the subject on me. She said she was trying to cut down on the stress in her life. She was cutting back her hours at work and taking time off to take her mother on some trips. We hung up, both looking forward to our lunch a week from Saturday.

On the Tuesday before I was supposed to see Josie, Pamela, who worked at Inner City Cultural Center with Josie, called me. She said, "Barbara, there's no easy way to tell you this ... Josie passed this last Friday." I couldn't translate the words I had just heard into reality, and could not respond. Pamela went on to tell me that Josie was supposed to see her mother that weekend but

186

didn't show up. Then that Monday, she didn't show up for work. They had never known her not to come to work without calling or letting someone know. When she didn't show up again the next day, a writer who worked at Inner City went over to her house to check on her. They got the landlord to open the door and they found her in bed. She was dead. The autopsy revealed that she had had a stroke and had been dead for several days.

Josie was my first adult friend and for many, many years my closest friend. She had been so dear and precious to me for so long that I couldn't believe that I'd never again be able to call her. Her memorable funeral was attended by the many people who worked with her and loved her. She had always been such a presence in the lives she touched that we were all in shock at her sudden death. I have wished for Josie so many times since then.

If there was going to be more tragedy in my life I certainly could not have imagined it. I felt too blessed. By now, my daughters were both college graduates and pursuing careers in their fields, Melodie in electrical engineering and Tiff in broadcast journalism. My life with Dave was full and satisfying. I didn't want anything to change, unless perhaps, there was a way to bring back the people in my life that I had lost.

John's mother, Hattie, had been in the hospital at UCLA several times. Dave and I went to visit her one Sunday, and we talked and laughed and had a great time with her. A couple of weeks after that John called. He was very upset. He said that his mother was very sick and we should try to get to the hospital that day. When we got there, she was in tremendous pain. We only saw her for a few minutes. She died a few hours later that same day.

At Hattie's funeral, the usher in the church, for some reason, seated me next to John. Dale was on the other side of him. When I sat down in the pew next to him he said, "God, I've got Barbara on one side of me and Dale on the other. I'd be better off up there in that casket with my mother." I don't know what he thought we were going to do to him.

Before the funeral began, Patty Cochran approached me and said, "You must be Barbara." Although we had been talking on the phone, this was the first time we had actually been in the same room together. It was a fitting symbol that we met at Hattie's funeral. By condoning her son's behavior and even covering for him, this otherwise dear woman had helped keep both of us in John's life far longer than either of us might have otherwise stayed.

On the Monday before Christmas in 1992, I was having my friend Mary over for lunch. Dave, who was a terrific cook, helped me prepare the food, but when Mary arrived, he said hello and talked for just a few minutes, then told us he was going to rest a while and eat later. Dave and I had planned to do some Christmas shopping after lunch, but when my friend left he said he was too tired and urged me to go on without him. He said he'd feel better tomorrow and would finish the shopping with me then.

The next morning he got up, got the paper, and came back to bed. He told me he just wanted to lie back down for a while until the house got warm. I said I'd get up and get the coffee. After I was up, the phone rang. I talked to my friend Gladys for a while and when the coffee was done, I told Gladys to hold on for a minute while I took Dave his coffee. When I went into the bedroom, Dave looked terrible. His face was flushed red. I rushed over to him. He looked up at me, trying to say something, but he couldn't talk. I ran back to the kitchen phone and hung up with Gladys. Then I ran back into the bedroom and called 911. Dave's face seemed to be getting redder. He kept straining as if trying to move. It was then that I realized he was paralyzed. Remembering the course my mother's last few days ran, I was filled with terror.

When the ambulance arrived, they took him to the nearest emergency hospital. I followed the ambulance to the hospital and when I got there called Melodie. When she arrived, I was sitting by Dave's bed, holding his hand. I said, "Dave, if you recognize Melodie, just squeeze my hand." He did. That was a good sign, I thought. Later, I tried to feed him and he was able to chew and swallow. This is another reason for optimism, I thought. He's going to be fine. I know he will.

At one point, he started to cry. My heart was breaking, but I tried to be strong for him. "You're going to be all right, Dave," I told him. I just wanted to go somewhere and weep, but I knew I couldn't. I stayed there for him. Later that afternoon the doctor asked to speak to me. He could not say the words I so much wanted to hear, that he saw something, anything, in the tests to give him hope that Dave would ever be able to walk or talk again. I couldn't believe what I was being told. "But what about with therapy?" I asked. He said he wasn't sure. He'd have to study the tests further.

They were supposed to transfer Dave to another hospital, but it frightened me even more when the doctor said they weren't going to transfer him that day. He was too weak to be moved.

I finally went home that night. The next morning when I went to the hospital, Dave didn't look as well as he had the day before. He could still squeeze my hand, but I could feel that he was losing strength. That day, I watched him slip in and out of a coma. By now I had called Dave's sons and they were there. I went home that evening and they promised to call if there was any change. Two friends came out to stay with me, Gladys and my friend Margaret, both dear wonderful women I had met teaching. Melodie also stayed at the house with me.

When I woke up the next morning, I thought I should call our minister, whom we affectionately called "Dr. Dan." He said he would meet me at the hospital. When I got to Dave's room, I was shocked at how he looked. He was now in deep coma, fully unresponsive. It seemed that the end was near, but I couldn't believe it. I loved him so much. I could not believe he was dying. The doctor came in and asked the family to wait in the waiting room, but allowed Dr. Dan to stay in with Dave.

When Dr. Dan came out of Dave's room, he was walking with his head down and his hands behind him. When he got to us he

189

asked us to pray with him. "Dear Lord, sometimes as loved ones we have our own selfish wishes and desires," he prayed, "but you as our Lord know what is best for one of your servants." When I heard his words I knew they were his way of telling us that there was no hope, preparing us for what he knew we would soon be facing. A few minutes later the doctor came out and said to me, "I did everything I could for David, but he's gone."

I went back into Dave's room to say good-bye. He was lying there very still and peaceful. I was in a state of shock. I hugged him and kissed him. All I could think was, no God, he can't be dead.

What dreams had died with that kind and good man! I had looked forward to growing old with him. We had talked about the places we planned to travel to together. Now, in a short few days all those dreams had turned to dust. Once again I had lost someone I loved dearly, and was left with the burden of going on in life without the company and affection of someone whose love I had grown to rely upon.

The next days were a blur for me. We went home and Melodie helped me make phone calls. People came by that evening to comfort me, including young Jonathan, Patty's sweet, sensitive son, now a college student. I think John also came over later that night. I know that he and Melodie went with me to the funeral home the next day to pick out the casket.

As the months went on and I mourned Dave's death, I had a lot of support. Melodie stayed with me for months. Friends arranged it so that I was never alone—someone came with an overnight case every weekend for the next three or four months. In the moments I did have to myself, I thought endlessly about Dave. More than anything, I was grateful that I had known him, even if our time together had been short. His sweetness and sensitivity had helped me risk loving again, had encouraged me take another chance on marriage—a state I had vowed I would never again enter. Because of Dave, I had known a level of happiness with a man that I had never thought possible. I thanked God that I had taken the risk.

Chapter 17

B
ack in 1968, the year O. J. Simpson won the Heisman Trophy, I remember thinking how terrific it was that a black guy had won it, especially one from my own state of California. Over the years, I watched him gain a second fame as an actor and celebrity, a personable guy with a ready smile and a confident, easy manner. In 1987, my daughter Tiffany Cochran and his daughter Arnelle Simpson were debutantes together at a cotillion put on by a black society club called "The Links."

The saying went that you knew you had made it when your daughter was invited to debut at a Links affair. John and I, then divorced, were both at Tiffany's cotillion, each with our new spouses. I remember seeing Arnelle Simpson and Marguerite Simpson, but neither Tiffany nor I remember O. J. being there.

Then, in June 1994, along with millions of other puzzled Americans, I watched O. J. Simpson and Al Cowlings take the L.A.P.D. on that bizarre Bronco chase. A few days earlier, when the news had broken about the murder of Nicole Brown Simpson and Ron Goldman, my first thoughts had been about Nicole's young children losing their mother at an even younger age than I had lost mine. Naturally, I hoped that O. J., a hero and role model to many blacks, had not killed the mother of his children.

But with my own history as an abused wife, this hope was diminished by what started to come out about the violence in O. J.'s relationship with Nicole in the years before her murder. In the days before the murders she had been trying to convince O. J. that there would be no more attempts at reconciliation.

From my own experience, as well as all I had learned about spousal abuse in the years since I made the break and left Johnnie Cochran, I knew that the period immediately after an abusive husband finally faces up to the fact that it is indeed all over can be a very dangerous time for an abused wife. And so, before there was any talk of DNA evidence, my own gut feeling was that if it turned out that O. J. was guilty, I would not be surprised. Of course, I had lived with a criminal lawyer long enough to know that people should be considered innocent until proven guilty.

At some point during the evening of that Bronco ride, or in the days that followed, as I watched Howard Weitzman speak to the press as O. J.'s lawyer, I asked myself, "Where's Johnnie?" for it seemed that here was a case calling out for Johnnie Cochran. I was surprised when the weeks went by and the ubiquitous TV cameras now showed Robert Shapiro at O. J.'s side—and still no John. John, as well as every other prominent criminal lawyer in the country, appeared as a commentator on one of the TV sideshows, giving his expert analysis on the case. But I knew John. He loved high-profile cases, and had always sought them out—from his first one, the Deadwyler case that propelled his career forward, to his most recent stint before the national media as the lawyer who negotiated the deal for Michael Jackson in which Jackson faced no criminal charges but paid some money to a young man who had charged him with sexual abuse.

A few days later, John happened to call and, in the course of conversation, I asked, "John, why aren't you defending O. J.?" He paused, then said, "Oh, you know, O. J.'s a friend." I smiled to myself. In the entertainment world John had entered in recent years, the word "friend" is used very loosely. "John," I said, trying to bring him down to earth, "this is Barbara you're talking to, not the press. You know who he is and he knows who you are, but what's this about friends?" But he only went into his deaf salesman routine, mumbling something like, "Oh, it's awfully hard to defend a friend. It would be a lot of pressure."

I dropped the subject. Less than a week later, Shapiro's office announced that John had been added to the "Dream Team," and soon thereafter he would become its captain. It wasn't long after John took over that this case became a case about race.

In the history of this country, unfortunately, there have been numerous false charges made by authorities against black male heroes—Malcolm X, Marcus Garvey, and Mohammed Ali come to mind as but a few examples. Their cases taught African Americans that in trials involving a black hero it was not unreasonable to continue to presume innocence, whether or not a court found the person guilty.

Cases like these have also imbedded in our black culture a deep-seated resentment against the "system" and the system's most aggressive and potent agent in the black community, the police. At any trial, after all, the main thing the judge in his robes and the lawyers in their suits and ties have to present to the jury is the evidence brought to them by the cops on the beat. History has taught the black community that many police officers see black males not as full citizens entitled to all the protections our constitution affords.

This is not to say that the African-American community is of one mind on this issue. I have heard blacks say, "Come on, now. Why would the police go about framing an innocent man?" But then, others might respond, "That's the whole idea. The police presume a black man is guilty. When they manufacture evidence, what they think they're doing is just making sure that a guilty man gets convicted."

Knowing full well the depths of these feelings, John, I'm sure, planned his defense of O. J., portraying him as a great black hero, an innocent man framed by racist cops who started out believing him guilty and who harbored an overpowering need to bring this high flyer down to size.

John once said, speaking about the Deadwyler case, "The issue of police abuse really galvanized the minority community." Now his hope seems to be to galvanize these latent feelings in the black members of this jury. If the attitudes of dismissed jurors Willie Craven and Jeanette Harris are indicative of those of others on the jury, then John's strategy may well prove successful. As John said to Patty Cochran shortly after he took the case, "If I can get just one black juror, then I can get a hung jury."

John's favorite motto, which is written in bold on his firm's brochure, is a quotation from Dr. Martin Luther King, Jr.: "Injustice anywhere is a threat to justice everywhere." But like many other things in John's life, this motto may be on the page solely to cultivate in others a certain image of him, not as an accurate description of any creed that he follows in his practice or his life.

He was never concerned about the injustice he did to me, to Patty, and to all of our children by living a double life for so many years. And though he may be very concerned about an injustice to O. J., he seems oblivious to the injustice that a defense strategy exploiting a deep-seated resentment among blacks may deny justice not only to Nicole Brown Simpson, Ronald Goldman, and their families, but to people everywhere who are concerned about positive race relations.

It is true that O. J. Simpson, a black man, deserves justice. But it is also true that Nicole Brown Simpson, a woman, and Ron Goldman, a white man, deserve justice. Justice also requires that racial antagonisms should not be inflamed in cases where there is no evidence that racism actually played a part in the arrest or the prosecution.

Viewing this case solely through the lens of race, as John hopes the jury will, or at least one juror will, cannot serve justice. It is the prosecution's job, not mine, to prove John's theory of a police conspiracy wrong. But I would like someone to address a central question for me. Why would all those white cops, and perhaps some black ones as well—who see O. J. as an amiable celebrity, ex-football hero, and movie star, whom they invite to their Christmas parties—risk their careers and their own freedom by planting evidence to frame him?

Unlike most other black heroes that have suffered harassment by police or prosecutors, O. J. has hardly been seen as promoting a political message that the white power structure might find challenging. He is no black power activist, far from a militant, surely no H. Rap Brown or Stokely Carmichael. When it comes to opposing racism, he is not even in a class with Jim Brown. And, unlike the Rodney Kings of the world, he is hardly a typical, powerless black ghetto dweller that a few rogue police officers might decide to pick on. Would those allegedly racist cops really risk getting caught and do jail time simply because O. J. had a white wife? Call me naive, but I just don't get it.

I do not mean to minimize the issue of police abuse and misconduct. It is the issue on which John built his career and it is a legitimate issue. The duty of police officers is to protect all citizens, not to mistreat selected citizens. A society where even a small portion of police officers are racist, where they treat whites and blacks differently, is by definition an unjust society. But, in my view, a society where men violently batter and abuse their wives, sometimes even killing them, and where a wealthy man with clever, high-powered attorneys can exploit the racial injustice perpetrated on others to help him get away with the murder of his wife, is also an unjust society.

If focusing on race is one part of John's strategy as head counsel, a second part clearly seems to be to fight as hard as necessary to keep wife-abuse testimony out of the trial. John has repeatedly characterized the beatings as "domestic discord." I remember how John once glossed over the abuse in our marriage, how he claimed to his mother and father that the violence against me was just "blown out of proportion." He told his father, "I didn't know that she was hurt" only a few days after he had held me and beat me in the head with his fists.

If O. J.'s problem at trial was going to be his history of spousal abuse, he had the right attorney. As O. J. watches himself dance around on videotape throwing jabs and making jokes about how if you have a need to punch your wife you can "just say you were working out," he can be confident that he has an experienced attorney who will do a convincing job of glossing over the significance of O. J.'s misogynist humor.

The prosecution, to provide the motive in their case against O. J., presented evidence that he fits the profile of a wife abuser. One of the cornerstones of that profile, Christopher Darden claimed, is the abusive man's need to control his wife. I read an article that quoted from Faye Resnick's book where she remembers Nicole Simpson saying, "O. J. always controls everyone and everything around him." Indeed, the story being told at the trial and in the press reveals not only that he has a need to exercise control, but that he has developed his own techniques for sustaining the control, often co-opting third parties whose influence might undermine his control. For instance, when he saw Ms. Resnick's relationship with Nicole as possibly facilitating Nicole's need to get free of him, he befriended Resnick and hired her close friend to act as a consultant on one of his exercise videos. When he decided Kato Kaelin's friendship with Nicole did not serve his interests, and no longer wanted him living at Nicole's, he didn't confront Kato and threaten him. Rather, he offered him free quarters on his own estate, thereby putting Kato in debt to O. J. and in fear that a wrong word to Nicole might cost him a plush deal. He tried to act similarly with Nicole's family, helping her father get established in business, and playing rich relative to all who'd accept it.

Again, I thought, O. J. has an attorney who has a lot in common with him. I, too, always felt John needed to control me, as well as all women around him. And his style too, was to start by trying to co-opt rather than confront. When, for instance, he wanted to make sure that April, Patty's daughter, did not create

196

problems for him, he went out of his way to befriend her, even trying to make her his ally in his disputes with her own mother.

Money, I learned from being married to John, is an insidious instrument of control, one that O. J. and John both used with great skill. Nicole Brown had met O. J. when she was only eighteen. From the first time they got together, he always wanted her with him, attending every football game he played. His demands on her time precluded her from getting the training necessary to support herself if she would ever need to do so in the future. I would guess that if she ever expressed any such inclination, O. J. assured her that there was no need, for he would always be there to take care of her.

Fortunately for me, when this issue arose in my relationship with Johnnie I stood my ground. By not succumbing to his pressure to quit my job, I successfully resisted his efforts to put me totally under his control. I somehow seemed to know that without an independent identity of my own, without my own income, he would use the inequality between us as leverage to bend me to his way. It must be said that I had certain advantages that all wives don't have: I was close in age to John and already had a profession when we married; and second, in working as a teacher I was fulfilling a lifelong dream, one that had been put in me by my mother, and one that I would not easily be talked out of.

As John became wealthier, he lavished on me the proper accoutrements for the wife of a man of his growing stature—the diamond rings, mink coats, big house, and fine cars—but he was stingy with his money when I asked for things I wanted. At first I believed that his attempts to deny me any discretionary spending was about his wanting to spend money only on things he thought were of value—that he didn't trust my judgment. I realize now that his stinginess was all about control. I laughed to myself when I read that in his opening statement in the trial, where he was trying to rebut the prosecutor's claim that O. J. was controlling, he made a comment about how when a wealthy person gives a gift, he doesn't expect anything in return. No, I thought, thinking back on my life with him, he doesn't expect anything—just that all future decisions will be taken with the giver's generosity in mind, and with the thought that the generosity can easily be withdrawn at the first sign of noncompliant behavior.

When I heard the testimony during the wife-abuse portion of the trial, about O. J.'s grabbing Nicole by the crotch and announcing to everyone present, "This is mine," I once again thought about my relationship with John. I realized that even today John, though perhaps not as crassly as O. J., still behaves as if he has a special claim on me. We both attended a banquet honoring outgoing Los Angeles Mayor Tom Bradley about a year ago. John came over to my table and announced, "I'm the only man here with two wives—Dale's over there and here's my wife, Barbara." Everybody laughed but me. Here I was, divorced from John for more than seventeen years, already the widow of another man, and to him I was still to be counted his wife. If my deceased husband, David, had been alive and there with me, it never would have occurred to me to announce that I was the only woman there with two husbands. I never felt I owned the man I was with—I simply choose to join with him.

John and O. J. would probably both defend their remarks by saying they were just kidding. But to me, their comments reflect the way they really feel about the women they are with. In the minds of men like O. J. and John, even divorce can't set a woman free.

When dismissed juror Jeanette Harris came on the local news and talked about how she felt sorry for O. J. and found Johnnie Cochran so "fascinating," I thought here is a woman who has bought the public images these men project and can't see beyond those images. Since this trial hit the airwaves, both my daughters and I have heard other women comment on how handsome or sexy or charming John was. Knowing him for as long and as well as I have, I was puzzled about what it is that women find so attractive in John. Then I watched a clip from the trial, in which John sidled up close to Marcia Clark, grinning and speaking softly to her. Even Prosecutor Clark, hardly a pushover for a snow job, had to return the smile. He has perfected the art of charming women. He certainly had lots of practice while we were married and even today, even in situations where he cannot expect to be successful in any attempt at seduction, as with Clark, he plies his craft, if for no reason other than to keep his skills sharp.

According to many press accounts, O. J. was also a womanizer while he was married. And like John, he would leave little bits of evidence of his outside conquests around—such as strange earrings in their bed—to let Nicole know other women found him irresistible.

I also found it enlightening that the flip side of philandering in abusive men is so often an intense jealousy of their wives. That certainly seemed to be true of both O. J. and John. Even if he is acquitted of murder, from the battering evidence we've all seen and heard, it cannot be denied that O. J. was violently jealous of Nicole. John always became furious if I received any attention from other men. If at a party I had received a compliment from a man, or some other social attention or special courtesy, I could expect some verbal abuse on the way home. And, of course, it was a phone call from a man, innocently calling me about joining a bowling league, that set off the most painful of the beatings that John ever gave me.

It seems to me that Orenthal James Simpson and Johnnie L. Cochran, Jr., are mirror images of each other in their apparent disdain for women. John finds it hard to show respect even to his peers when they are women. He recently slipped in court and showed his true colors when he referred to Marcia Clark as "hysterical." I was delighted when she immediately protested his sexist remark. When he then said the term had nothing to do with women and therefore could not be considered sexist, Judge Ito told him to look up the word "hysterical" in the dictionary; its root is from the Greek for womb.

John is an old-fashioned man, who can't conceive of men and women as equals. He called on the phone about a year or so ago and told me of a dinner he and Dale had with a woman who had recently been appointed president of a local university and her husband. In total disbelief, he said, "Barbara, her husband left his job in order to move out here so that she could be president of that university. Can you believe that?" Many men have become more open-minded and progressive about equality of opportunity for the sexes over the past couple of decades, but not John.

John and O. J. are two middle-aged stars who have used their considerable ambition and skill to achieve both high praise in their respective fields and financial success. But behind their smooth, charismatic exteriors, I can't help but see two men who have very little respect for women, who abuse and need to control the women in their lives, who use their money as a means of control, and who are routinely unfaithful to their wives.

Given his own deeply sexist attitudes and behavior, it is not surprising to me that as O. J.'s attorney John would focus this case on race, using all his considerable skill to gloss over the very real and critical issue of O. J.'s history of physical and emotional abuse of his wife.

Did O. J. kill Nicole? I'm sure John has never asked O. J. to answer that question, because the answer might create ethical problems for John. And to lawyers, the term "ethical problems" does not describe those problems having to do with conscience, but only those that the lawyer believes might land him before the ethics committee of the local bar association. But in O. J.'s case, I would guess that John's defense strategy was probably mapped out before he ever received the call to join the team. To John, as to too many defense lawyers, the important factor is not what is true, but what the prosecution can prove, and what shreds of doubt can be introduced to undermine the prosecution's offers of proof. As he sees his job, he must create a picture of the case as the biggest police wrongdoing case he's ever had. That women across this country, black women and white women, see the case as much more, is of no interest to him.

My hope is that the jury will scrutinize John's claims more carefully than either Patty or I did when we heard the masterfully fabricated tales he told us. If the evidence against O. J. is enough to convict another man, an ordinary man, black or white, then it should be enough to convict him. That is the standard I hope the all the jurors hold to. Otherwise, this trial of the century will be an injustice to the victims and, as Dr. King said of all injustice, will be a threat to justice everywhere.

Chapter 18

L.A.'s black community recently inducted Johnnie Cochran into its "Promenade of Prominence," a walkway that honors people they see as role models in the community. At the award ceremony, John said to the audience of mostly blacks, "I want to carry myself in such a way that you will be proud and you will want to go out and seek and find an African-American lawyer or doctor."

I heard it as one more example of the dichotomy in John's onstage/backstage dual personality. I hope this book causes some people who know only the onstage Johnnie Cochran to question if men like him are really the kind of people we in the black community should hold up as role models for our children. The institutions of family and marriage were seriously damaged among African Americans during the centuries of slavery, when slave traders and owners sold families apart. Later on, these important institutions were further undermined when work opportunities for black women were for the most part restricted to acting as servants in white homes, while most black men weren't even afforded that level of work. It is important that when our community defines people as role models we take these historical realities into account and chose as models those who will encourage a return to the strong family values African Americans brought to this continent and had stripped from them here.

Do we really want our children trying to emulate those who philander, exploit wives and mistresses alike, impregnate women who are not their wives, and who emotionally, verbally, and even physically abuse the women in their lives?

One of the moderators at that same award ceremony said, "They messed around with Michael Jackson, they messed around with Mike Tyson, but they better not try to mess around with Johnnie Cochran because Johnnie Cochran is a true role model. I mean you might find a little something on these other folks, but don't mess with Johnnie Cochran."

I have "messed" with Johnnie Cochran, but only because he messed with me for seventeen years, and because bearing witness and setting the record straight is an important part of the healing process for all those who have been exploited. As well, I don't think our black role models and heroes should be exempt from careful scrutiny, any more than role models of any other race or ethnicity should be exempt.

An African-American writer, Clarence Page, wrote in a recent article in the *Los Angeles Times* that we must hold our black role models up to "the cleansing fire of the highest ethical and moral standards. We cannot demand much better of others if we do not weed our own garden first." To shift the metaphor, let me go to words I once heard on the subject of housekeeping: "Dirt swept under the rug does not disappear. It just grows into a larger and larger hidden pile, till one day you trip on it."

So to those who say to me, "Why are you bringing that man down," I say I am not. If this book lowers him a notch (there is nothing I can imagine that would bring him down), it will be his own actions, and the truth told about those actions, that will have done it. I for one will no longer stand aside passively and deny what he has done—for doing so condones such behavior, not only in his life but in the lives of others who see a certain macho style in abusing the women in their lives.

Martin Luther King, Jr., said that, "Freedom is never voluntarily given by the oppressor; it must be demanded by the oppressed." In writing this book, I have celebrated my decision eighteen years ago to demand my freedom. It is my hope that all women who have silently accepted mistreatment at the hands of a man with power, prominence, wealth, or emotional influence over them will also be inspired to demand their freedom. The message I wanted to convey is not only that I was part of a pattern of abuse that I now know is widespread in America, but that there is a way out.

We have long known that it is important for all women in abusive relationships to know that they are not alone out there, that there are others who share their experience. But even more important is that these women understand that such tortured relationships need not be life sentences, that they can break free and build new lives for themselves. If this book encourages even one woman to try, it will have been worth all the flak I know will be thrown at me for having undertaken it.

We women have enormous power when we have the courage to use it. Where would the Civil Rights Movement have been without Rosa Parks, who struck the spark igniting that powerful and far-reaching peaceful revolution? In my classroom, when I read Rosa Parks's story to my first graders, there's usually not a sound in the room because, even as young as they are, they understand courage.

In the mostly white school where I teach today, I often don't have a black child in my class, so when we act out the story, I usually have to choose a little white girl to play Rosa Parks. I explain to the class that she's going to play a black lady and they have to imagine that. A little girl named Elizabeth played Rosa Parks one year and she was just ferocious. When the bus driver approached her and asked her to move to the back of the bus, she cried out, "No! I'm not getting up out of my seat for a white man!"

Elizabeth so well captured the righteous anger of the moment that tears rolled down my cheeks, all the way down to my neck, ruining a good blouse. This tableau, with little Elizabeth as Mrs. Parks, taught that class of white children not only something about the Jim Crow South, but also taught them something about the universality of the human need for justice.

Women have to demand fair treatment just as surely and just as unequivocally as Rosa Parks did. A respect for tradition does not require that women who have been humiliated and abused should stay in an abusive situation for any man—black or white. Or, if they don't, be branded as malcontents or troublemakers. Or even as man-haters.

The process of writing this book has not been easy for me. I have had to recall events in my life that were very painful, events I had long buried in the deepest vaults of my memory. I have had to face up to the fact that people who were close to me, people I trustingly and blindly believed in, betrayed me, out of fear of or loyalty to John.

I have also had to look at myself, my own weaknesses and strengths. From the innocent, naive young woman I was when I married John, I have grown enormously. My youthful enthusiasm has been replaced by deeper and more durable emotions, and I am a much stronger woman now.

Writing this book, taking my freedom, if you will, from my painful and degrading past with John, has reinforced the strength I have been rebuilding since I left him. I feel a sense of relief as this process ends, a sense of healing.

My minister counseled me before I began, saying we all need to "empty ourselves out in order to be fulfilled." I have. And with the past no longer disguised in John's tangled web of lies, I feel cleansed, renewed, and truly "free at last."

I am also optimistic for the future—for both myself and the women I hope this book may touch in some way. For myself, I am grateful for the good people in my life—my children and all my dear wonderful friends. I am even hopeful that in the years to come there may be someone out there, someone who is seeking love and companionship from a woman he is prepared to accept as a partner, not as a piece of property, and that I may once again experience the kind of close, loving relationship that I learned was possible through my relationship with my late husband, David.

I am also optimistic for women. I hope that women who have read my story will look at their own lives. As my life with John illustrates, there are many forms of abuse. A woman doesn't have to have visible bruises or bleeding to be suffering tremendously.

My hope is that other women will value themselves enough to feel that they deserve to be treated with nothing less than a full measure of love and respect.

The bond of womanhood is strong. As women, we need to help each other develop our strength and courage, rather than teach each other to be content with dependence and fear. We need to set independent goals for ourselves, and not remain dependent on the men in our lives for our meaning and identity. We need to go after our own successes, our own achievements, so that no matter what happens in our relationships or our marriages, no matter what a man may say or do, we never surrender to another the power to validate us as persons of worth.

I fully expect my ex-husband, the father of my children, Johnnie Cochran, to attack me in whatever way he can for writing this book. I expect him to impute the basest of motives to me, and to sneer at my values and accomplishments, just as he would work to undermine the credibility of any witness for the prosecution in a case he is trying.

But now he will be attacking me on a more level playing field. Not that I have the resources to bring to a battle that he has. I don't. But solely because I no longer have to defend my actions in terms of whether or not they represent proper behavior for the wife of Johnnie L. Cochran, Jr.

Now I know enough to defend my life in terms of whether or not it has responsibly and fairly served the needs of my children, my family, my friends, my students, my society, and certainly myself, not on the basis of whether or not it adequately served my husband's ambitions.

It will always be tough for women to stand up to men like John. Such men seem constitutionally incapable of accepting responsibility for what they have done. The words, "I'm sorry. I shouldn't have done all that," cannot pass their lips, probably because such words are barred from their conscious thoughts. Well, in John's case, whether he can face it or not, he definitely shouldn't have done all that. But he did. And now, as they say in his line of work, "Let the record so state."

If you or someone you care about is being abused, call any of these numbers for support, information, and referrals.

National

NATIONAL RESOURCE CENTER ON DOMESTIC VIOLENCE: 800–537–2238

BATTERED WOMEN'S JUSTICE PROJECT: 800–903–0111

NATIONAL COALITION AGAINST DOMESTIC VIOLENCE: 303–839–1852 (in Colorado); 202–638–6388 (in Washington, D.C.)

NATIONAL BATTERED WOMEN'S LAW PROJECT: 212–674–8200

NATIONAL DOMESTIC VIOLENCE HOTLINE: 800–333–SAFE

In Los Angeles

LOS ANGELES RAPE AND BATTERING HOTLINE: 310–626–3393